Morphological Basis of Human Reproductive Function

Edited by

Giovanni Spera
Professor of Andrology
University «La Sapienza» of Rome
Rome, Italy

and

David M. de Kretser
Chairman, Department of Anatomy
Monash University
Clayton Victoria, Australia

Springer Science+Business Media, LLC

Proceedings of the 1st International Symposium on Morphological Basis of Human Reproductive Function, held September 11-12, 1986, in Fiuggi Terme, Italy

© 1987 ACTA MEDICA EDIZIONI E CONGRESSI s.r.l.
Piazza Montegrappa, 4
00195 Rome, Italy

ISBN 978-1-4612-9083-4   ISBN 978-1-4613-1953-5 (eBook)
DOI 10.1007/978-1-4613-1953-5

CONTENTS

OVARY

EPIDYDIMIS AND SPERMATOZOA

Preface

Reproduction is, it would appear, a very simple biological event: the result of an act of love. In actual fact, it has always been, and still continues to be, one of the most complex, yet at the same time, most fascinating, problems, with which Science has ever been engaged.

Physiopathology of human reproduction has always made use of investigations of a morphological nature, perhaps on account of the need to «see inside» the reproductive apparatus or within the gonads themselves in order to better understand how they function. Observation of spermatozoa practically coincided with the discovery of the microscope, and histological study of the testis and ovary was, for a very long time, the only means available with which to closely follow the evolution of gametes destined to their meeting.

Improvement of techniques resulting from the evolution of endocrinology of reproduction has only apparently put aside morphological techniques, whilst at the same time there has been a gradual development of ultrastructural techniques, on the one hand, and, on the other, macroscopic diagnostic systems through images.

It is some time, not only since the state of the art was ascertained, but also since a co-ordinated review was made of all the morphological techniques available for the study of human fertility: in this volume, emerging from the Proceedings at the First International Symposium on the Morphological Basis of Human Reproductive Function, held in Fiuggi Terme, Italy, 11-12 September 1986, our aim was, with the enthusiastic agreement of the extremely qualified authors of the various chapters, to focus attention on the different aspects of reproduction from a single viewpoint: the form, microscopic, ultrastructural, histochemical, but also direct and indirect appearance of the gonads, accessories and gametes, with respect to their function.

Judging from the images proposed, the result has been spectacular, but perhaps even more exciting are the concepts of fundamental physiopathology triggered by these images, in support of those who, like us have always believed in this morphological approach to the problem and to the benefit of those biological and clinical research workers, will find in this volume the most up-to-date indications and knowledge through a rapid consultation of the pictures.

A special word of thanks to Ente Fiuggi S.p.A. without whom the meeting would not have been posible and to Acta Medica Edizioni e Congressi for organizing the publication of this volume.

G. Spera   **D.M. de Kretser**

May, 1987

An overview of functional morphology of human testis

A. FABBRINI, S. FRANCAVILLA, M. MARTINI,
G. PROPERZI, V. SALFI, P. ROSATI, V. SANTIEMMA

Clinica Medica Generale, Università dell'Aquila, L'Aquila, Italy

In the field of reproduction, a number of physiological
and pathophysiological problems remain to be solved
which are not explained solely by the secretory characteristics of the gonadotropins, prolactin and steroids.
In fact, our knowledge of the detailed control of
male fertility is poor; this is reflected in the
virtual absence of an effective therapy for infertile
men.

In any organ or tissue containing multiple cell types,
local coordination of the functions of the different
cells is fundamental to the efficient workings of
the organ. This coordination is realized through
paracrine and autocrine mechanisms. An improved knowledge of the local regulation of cellular functions
could help in clarifying some of the numerous, still
poorly understood, pathophysiological aspects of clinical
endocrinology.

In recent years, numerous investigations addressed
to the paracrine control of spermatogenesis and testosterone production.

The leading idea which stems from these studies is
that the biological effects of the two main extratesti-
cular trophic hormones, FSH and LH, on testicular
targets, are largely modulated by local factors (1).
Let us take as a paradigmatic example the steroidogenetic function of the testis. LH is required to maintain
the morphofunctional integrity of the Leydig cell,
namely the synthesis of testosterone (2). This steroid
is essential for the completion of spermatogenesis
through its stimulatory action on the Sertoli cell
(3). The seminiferous tubule seems to modulate the
effect of LH on Leydig cells through factors which

1

are produced by Sertoli cells (1). In this way the testosterone syntesis is optimized and adapted to the local need of the steroid, which is variable throughout the cycle of the seminiferous epithelium. The need for testosterone is high during the stage VII, VIII of the cycle in the rat (late spermatid maturation and spermiation, and appearance of prelepto- tene spermatocytes) and decreases thereafter (4). This different need for testosterone cannot be achieved by a central acting mechanism, but by a local communi- cation between the seminiferous tubule and the inter- stitium. The existence of a bidirectional interaction between the seminiferous tubule and the interstitium helps also to explain the different morphologic appea- rance of Leydig cells.

In the rat, Bergh (5) demonstrated that Leydig cells are larger around the tubules at VII, VII stages of the cycle when infact the need for testosterone is highest. In human it is much more difficult to demon- strate a cyclicity of Leydig cells in relation to the stages of the seminiferous epithelium, due to the small size of each stage along the tubules. However, the frequent observation in human testicular biopsies of Leydig cells with different cytological appearance, inthe same interstitial area suggests that this could be due to different messages coming from neighbour seminiferous tubules (Fig.1).

A stimulatory effect on Leydig cells which is originated in the seminiferous tubule seems to be greatly enhan- ced in cases of deranged spermatogenesis. This was suggested already in 1978 by Aoki and Fawcett (6) who documented a local hypertrophy and hyperplasy

of Leydig cells in proximity of seminiferous tubules where the spermatogenesis was deranged by local implants of cyproterone acetate. The hyperstimulation of Leydig cells was not detected far from the implants of the antiandrogen where the spermatogenesis was unaffected. This effect could be explained only claiming the existence of a local stimulatory mechanism origi- nating from the seminiferous tubule. These same inter- actions between the interstitium and the seminiferous tubule seem to be operative also in man, and help to explain data we obtained in some investigations on infertile patients.

Testicular biopsies from severe oligozoospermic patients showed the ultrastructural aspects of Leydig cells

2

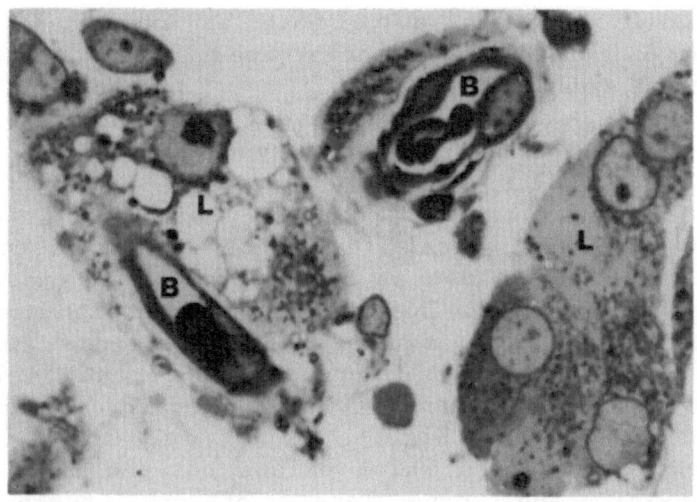

Fig.1.
B: Blood vassel
L: Leydig cells. The Leydig cell on the left shows
   voluminous cytoplasmic vacuoles.
X 1000 (original magnification).

damages (7); the quantitative analysis of the interstitial cells in the same biopsies revealed a Leydig
cell hyperplasia, and Leydig cell density resulted
positively correlated with both LH and FSH serum
levels (8 , 9).
The direct correlation between the number of Leydig
cells and serum level of LH suggested that the increased
density of interstitial cells could be related to
an exagerated stimulation by the gonadotropin. However,
in man with normal testicular function, HCG treatment
does notlead to any change of the Leydig cell density
(10). A true Leydig cell hyperplasia is,on the contrary,
observed both in man (Klinefelter syndrome) and animals,
when the spermatogenesis is greatly deranged (1).
In rodents , it has been recently found but not yet
characterized, a polypeptide factor produced by the
seminiferous tubule, which enhances the biological
effects of LH on Leydig cells; this factor increases
in the interstitial fluid, after treatments which
lower the interstitial levels of testosterone or
which induce a primitive seminiferous tubule damage
(11), (12). These findings suggest that the observed
Leydig cell hyperplasia in patients with a poor sperm

3

count, could be related to both an increased level
of LH and of a local factor originated in the damaged
seminiferous tubules.

In conclusion, a primitive Leydig cell defect could
be responsible for a decreased synthesis of testosterone; this results in: (A) increased level of LH,
and (B) derangement of spermatogenesis which infact
requires very high local levels of the steroid. The
latter effect has as a consequence, an exagerated
activity of the local positive feedback which through
a paracrine mechanism tunes the testosterone release
to the need of the seminiferous tubule, and in cases
of a Leydig cell damage, tries to restore a proper
testosterone synthesis(Fig.2).

The direct correlation which exists between the Leydig
cell density, and serum levels of FSH, suggests a
sertolian origin of this local factor.

To better understand the relationship which exists
between the interstitium and the seminiferous tubule
in man, we performed a quantitative evaluation of
spermatogenesis in groups of patients selected on
the basis of their Leydig cell density.

In each patient, 20 cross sections of seminiferous
tubules were evaluated by counting the different
types of germ cells including: spermatogonial stem
cells (Ad), activated spermatogonia (Ap+B), primary
spermatocytes (SC), round and elongating spermatids
(Sa+Sb), spermatids in maturation phase (Sc+Sd).
Patients were divided into three groups according
to Leydig cell density:

Group A: highest Leydig cell density.

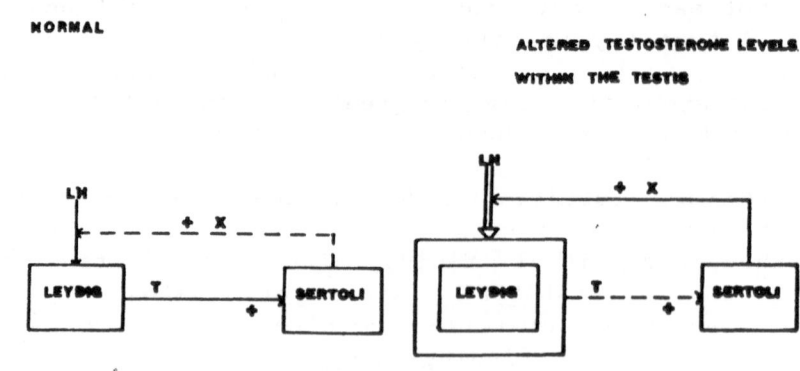

Fig.2.

Group B: intermediate Leydig cell density.
Group C: low (normal) Leydig cell density.
A Leydig cell hyperplasia resulted associated to an increased number of spermatogonial stem cells, and a lower number of mature spermatogonia, primary spermatocytes and spermatids, mainly late spermatids (Table I).
The results suggested that a Leydig cell hyperplasia is associated to a lower productivity of spermatogenesis. The productivity of spermatogenesis depends on:
A-the rate of differentiation of spermatogonial stem

Table I. Quantitative analysis of spermatogenesis. Patients selected on the basis of the Leydig cell density

GROUP		$\frac{Ly}{T}$	$\frac{Ad}{T}$	$\frac{(Ap+B)}{T}$	$\frac{SC}{T}$	$\frac{(Sa+Sb)}{T}$	$\frac{(Sc+Sd)}{T}$
A	M	$9.13^{(a)}$	$8.62^{(b)}$	13.78	36.9	5.58	9.12
	DS	0.75	2.38	1.71	8.90	2.16	7.27
B	M	$6.37^{(a)}$	6.07	13.47	35.03	7.38	9.13
	DS	0.77	1.59	2.98	10.66	3.48	6.67
C	M	3.06	7.45	17.13	46.85	11.35	$12.58^{(c)}$
	DS	1.00	1.19	4.43	14.07	7.86	5.18

T: Tubule cross section; Ly: Leydig cell; Ad: Staminal or A dark spermatogonia; Ap: (A pale) Spermatogonia; B: Bspermatogonia; SC: Spermatocytes; Sa-d: young (Sa, Sb) and mature (Sc,Sd) spermatids.
a: $P < 0.01$ between LY/T in group A and B and in group B and C.
b: Ad/T in group A is significantly higher with respect to group B and group C ($P < 0.05$).
c: (Sc+Sd)/T in group C is significantly higher with respect to group A and B ($P < 0.05$).
M: Mean
SD: Standard Deviation

cells in mature spermatogonia;
B-the rate of proliferation of spermatogonia and
commitment into the spermatogenic process (appearance
of preleptotene spermatocytes);
C-completion of first and second meiotic division;
D-spermatid maturation;
E-cell degeneration. The number of mature spermatids
is far smaller than expected due to degeneration
of germ cells. This is genetically determined in
each species and peaks during the proliferation of
mature spermatogonia, early stages of meiosis, and
early stage of spermatid maturation (13).
The low productivity of spermatogenesis in the patients
investigated resulted associated to:
A- increased cell loss at level of mature spermatogonia,
primary spermatocytes and spermatids, with a peak
of loss of late spermatids;
B- apparent compensatory increase of number of spermatogonial stem cells.
These results are very similar to those observed
in rodents following pituitary ablation (14 , 15).
The germ cell loss at specific stage of development,
in the operated rats is largely prevented by supplementation with LH or testosterone propionate, and
completely corrected after supplementation with both
LH and FSH, FSH alone having no effect. In particular
testosterone or LH supplementation prevents degeneration
of mature spermatogonia, protects spermatocyte development (being required for the transformation of
zigotene spermatocytes to pachitene spermatocytes)
and fully supports complete spermiogenesis (14 , 15).
All these data support the hypothesis that the altered
spermatogenesis in the investigated patients showing
an increased Leydig cell density could be due to
an insufficient supply of testosterone.
The physiological role of testosterone on the seminiferous tubule is mediated by Sertoli cell which is the
only tubule target for the androgen (3).
Sertoli cell creates a special environment through
its special intercellular junctions, which seems
to be essential for the differentiation of the germ
cell (16). Sertoli cell produces , moreover , a large
number of more or less identified substances whose
role in maintaining the maturation of germ cells
is, in most of the cases ,just supposed but not completely
demonstrated. An unquestionable way by which Sertoli

6

cell supports germ cells differentiation, consists in supplementing germ cells in the adluminal compartments of the tubule, of proper energy substrates.

During meiosis and spermiogenesis, germ cells undergo peculiar modifications of their mitochondria, which are the morphological expression of a metabolic change consisting in the passage from a slow or normal respiration to a high respiratory activity (17).

Lactate and pyruvate are the essential energy substrate to maintain the high respiratory activity of spermatocytes and spermatids (18). Sertoli cells are the likely source for supply of these substrates because these cells carry out glycolysis at a high rate and secrete lactate under aerobic conditions, utilizing a large amount of glucose, which on the contrary, does not seem to be a proper substrate for germ cells (19). In a recent elegant investigation, Jutte et al. (20) demonstrated that pachitene spermatocytes and round spermatids which in culture do not survive in absence of Sertoli cells, are able to maintain for a while , a specific pattern of protein synthesis in the absence of Sertoli cells but in presence of a proper lactate and pyruvate supplementation.

These observations certainly emphasize the role of "nurse" cells played by the Sertoli cell.

Experiments recently performed in our laboratory demonstrated that cultured Sertoli cells harvested from pubertal rats have a high activity of LDH, the enzyme which catalyzes the interconversion of pyruvate and lactate.

Sertoli cells have all five LDH isoforms, and hormone stimulation of the cultured cells is responsible for a shift of the enzyme toward the higher capacity isoform and for an enhanced activity (21). Our data are consistent with glucose oxidation by the Sertoli cell being preferentially oriented toward lactate production, rather than ending in the citric acid cycle.

Germ cells can utilize the energy provided by the Sertoli cell as lactate via its conversion to pyruvate by LDH-X, the citric acid cycle, and the respiratory chain (Fig.3).

All these data suggest that a low quantitative efficiency of spermatogenesis may be a consequence of a limited availability of the energy substrates to the germ cells in the tubular environment, due to

7

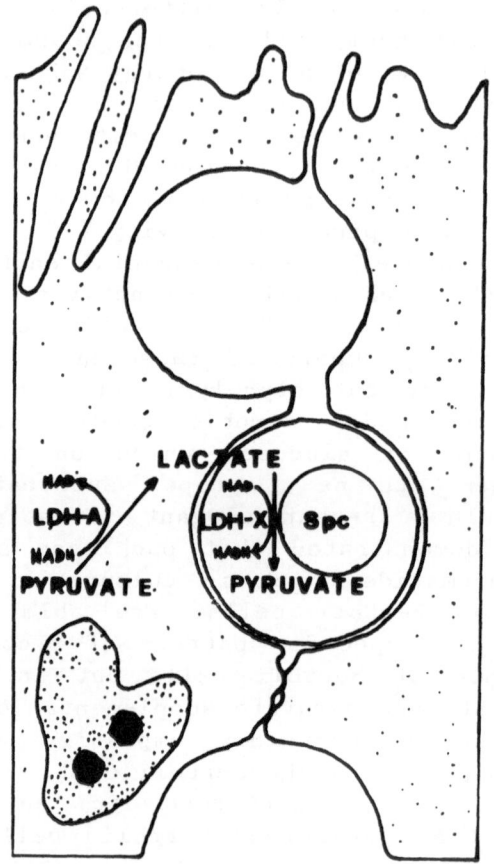

Fig.3

a decreased energetic metabolism of the Sertoli cell.
A hormonal imbalance, by affecting the Sertoli cell,
may be responsible for a prompt reduction or disappearance of spermatocytes and spermatids which are the
germ cells more directly dependent on the availability
of the energy substrates provided by Sertoli cells.

References
1-Sharpe RM. Paracrine control of the testis. Clinics
    Endocrinol. Metab. 1986: 15; 185-207.
2-Dym M., and Madhwa Raj HG. Response of adult rat
    Sertoli cells and Leydig cells to depletion of
    luteinizing hormone and testosterone. Biol. Reprod.
    1977: 17; 676-696.

8

3-Hansson C., Weddington S.C., Mc Lean W.S., Smith A.A., Neyfeh S.N., French F.S., and Ritzen E.M. Regulation of seminiferous tubular function by FSH and androgen. J. Reprod. Fertil. 1975: 44; 363-375.

4-Parvinen M. Regulation of the seminiferous epithelium. Endocr. Reviews. 1982: 3; 404-417.

5-Bergh A. Local differences in Leydig cell morphology in the adult rat testis: evidence for a local control Leydig cells by adjacent seminiferous tubules. Int. J. Androl. 1982: 5; 325-330.

6-Aoky A. and Fawcett D.W. Is there a local feed-back from the seminiferous tubules affecting activity of the Leydig cells? Biol. Reprod. 1978: 19; 144-158.

7-Fabbrini A., Santiemma V., Francavilla S., Moscardelli S., Francavilla F., Incorvati N., Bellocci M., and De Martino C. Leydig cell morphology and function in varicocele. In: Oligozoospermia-Recent Progress in Andrology. eds. G. Frajese, E.S.E. Hafez, C. Conti, and A. Fabbrini. Raven Press 1981: 77-94.

9- Francavilla S., Bruno B., Martini M., Moscardelli S., Properzi G., Francavilla F., Santiemma V., and Fabbrini A. Quantitative evaluation of Leydig cells in testicular biopsies of men with varicocele. Arch. Androl. 1986: 16; 111-117.

10-Heller C.G., and Leach D.R. Quantification of Leydig cells and measurement of Leydig cell size following administration of human chorionic gonado-tropin to normal men. J. Reprod. Fertil. 1971: 25; 185-192.

11-Sharpe R.M., Kerr J.B., Fraser H.M., and Bartlett J.M.S. Intratesticular factors and testosterone secretion. Effect of treatments that after the level of testosterone within the testis. J. Androl. 1986: 77; 239-245.

12-Risbridger G.P., Jenkin G., and De Kretser D.M. The interaction of HCG, Hydroxysteroids and intersti-tial fluid of rat Leydig cell steroidogenesis in vitro. J. Reprod. Fert. 1986: 77; 239-245.

13-Roosen-Runge E.C. The process of spermatogenesis in animals. Cambridge University Press London.1977.

14-Russel L.D., and Clermont Y. Degeneration of germ cells in normal, hypophysectomized and hormone treated hypophysectomized rats. Anat. Rec. 1977: 187; 347-366.

15-Chowdhury A.K. Dependence of testicular germ cells on hormones: a quantitative study in hypophysectomized testosterone- treated rats. J. Endocr. (Great Britain) 1979: 82; 331-340.

16-Fawcett D.W., Leak L.V., and Heidger P.M. Electron Microscopic observations on the structural components of the blood-testis barrier. J. Reprod. Fert. 1970 Suppl.10; 105-122.

17-De Martino C., Floridi A., Marcante M.L., Malorni W., Scorza-Barcellona P., Bellocci M., and Silvestrini B, Morphological, histochemical and biochemical studies on germ cell mitochondria of normal rats. Cell Tiss. Res. 1979: 196; 1-22.

18-Jutte N.H.P.M., Grotegoed J.A., Rommerts F.F.G., and Van der Molen H.J. Exogenous lactate is essential for metabolic activities in isolated rat spermatocytes and spermatids. J. Reprod. Fert. 1981: 62; 399-405.

19-Nakamura M., Fujiwara A., Yasumasu I., Okinga S., and Arai K. Regulation of glucose metabolism by adenine nucleotides in round spermatids from rat testis. J. Biol. Chem. 1982: 257; 13945-13950.

20-Jutte N.H.P.M. Jansen R., Grotegoed J.A., Rommerts F.F.G., and Van der Molen H.J. Protein synthesis by isolated pachytene spermatocytes in the absence of Sertoli cells. J. Exp. Zool. 1985: 233; 285-290.

21-Santiemma V., Salfi V., Casasanta N., and Fabbrini A. Lactate dehydrogenase and malate dehydrogenase of rat Sertoli cells. IV European Workshop on molecular and cellular endocrinology of the testis. Capri 1986, C 9.

New observations on the development of Leydig cells and the paracrine regulation of Leydig cells by the seminiferous tubule

D.M. DE KRETSER, A.E. JACKSON, P.C. O'LEARY, G.P. RISBRIDGER

Department of Anatomy, Monash University, Melbourne, Victoria, Australia

Two aspects of Leydig biology have received considerable attention in recent years, namely the cellular source from which Leydig cells arise and the local control of Leydig cells by the seminiferous tubules. New data concerning both aspects of Leydig cell biology have been derived recently from the use of the agent, ethane dimethane sulphonate (EDS), which had been shown previously to destroy Leydig cells (Cooper and Jackson 1970; Morris and McClukie 1979). This paper reviews the recent literature providing new observations from our work and that of other investigators concerning these aspects of Leydig cell biology.

The Cellular Source of Leydig Cells

There is in general agreement that a foetal generation of Leydig cells develops at the time of sexual differentiation in the male and subsequently regresses to leave a testis devoid of Leydig cells in the prepubertal period (1,2) Furthermore, the onset of sexual maturation during puberty is associated with the development of a second generation of Leydig cells which persists throughout adult life (1). However in some species such as the rat and the pig, the two phases are not sufficiently separated in time to permit a clear distinction between the two generations of Leydig cells (3,4).

There is general agreement that the foetal generation arises from mesenchymal precursors of the gonadal ridge since cells intermediate between mesenchymal cells and Leydig cells can be frequently observed during this phase (2-6). However the cellular origin of the adult generation is less clear and the question of whether differentiation of precursors to Leydig cells occurs throughout adult life has not been resolved. Some investigators have suggested that the fibroblast-like cells, often loosely termed mesenchymal cells, are the source of the adult generation of Leydig cells (1,7,8). Others have suggested that the macrophages may be the source (9) or alternatively that the foetal generation may differentiate and provide a pool of cells for subsequent redifferentiation (10).

11

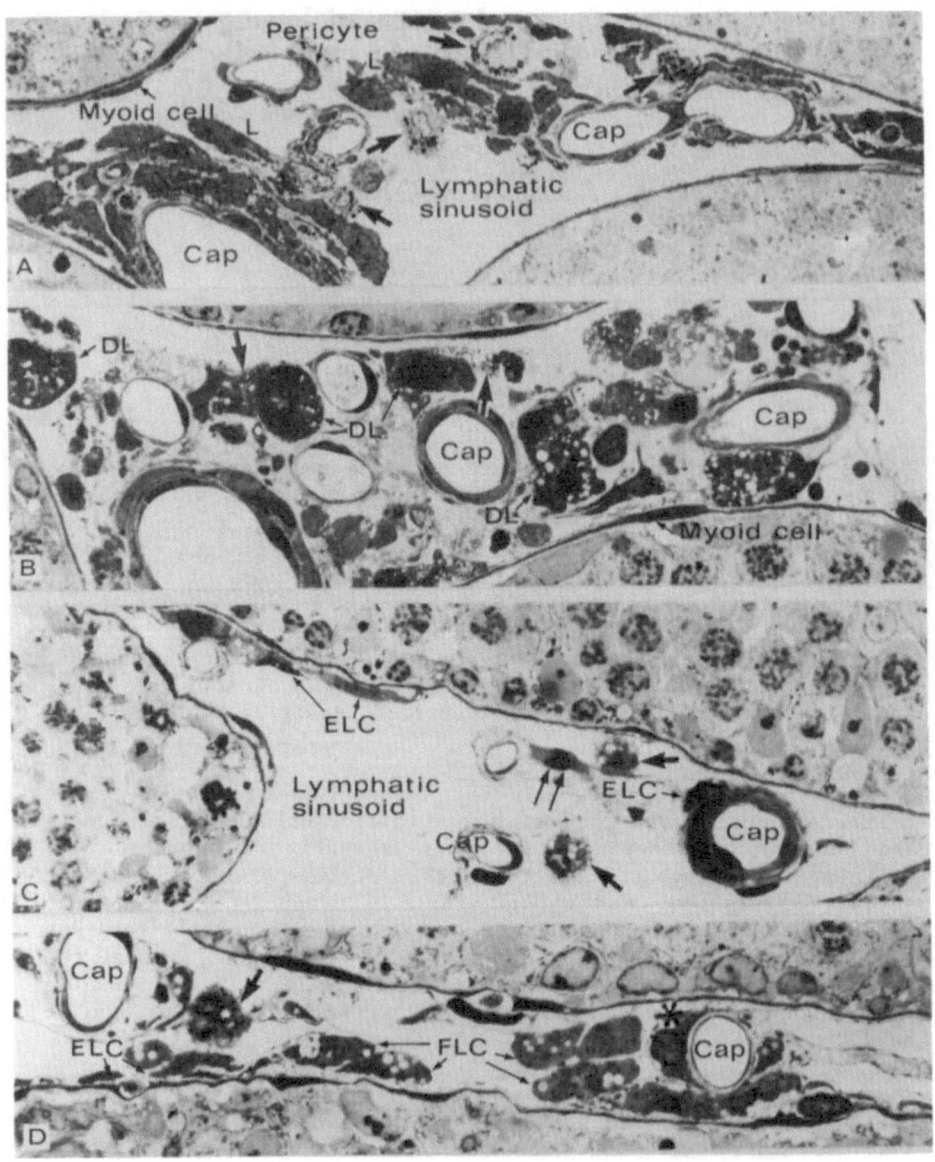

FIG. 1. These light micrographs from rat testes shows the
intertubular area. (A) normal rats (B) 24 hours post EDS (C) 21
days post-EDS (D) 35 days post-EDS). Abbreviations: - CAP =
capillaries, L = Leydig cells, ELC = Elongated Leydig cells, FLC =
Foetal Leydig cells, DL = Degenerating Leydig cells, Macrophages
denoted by arrows and a dividing Leydig cell precursor by double
arrows.

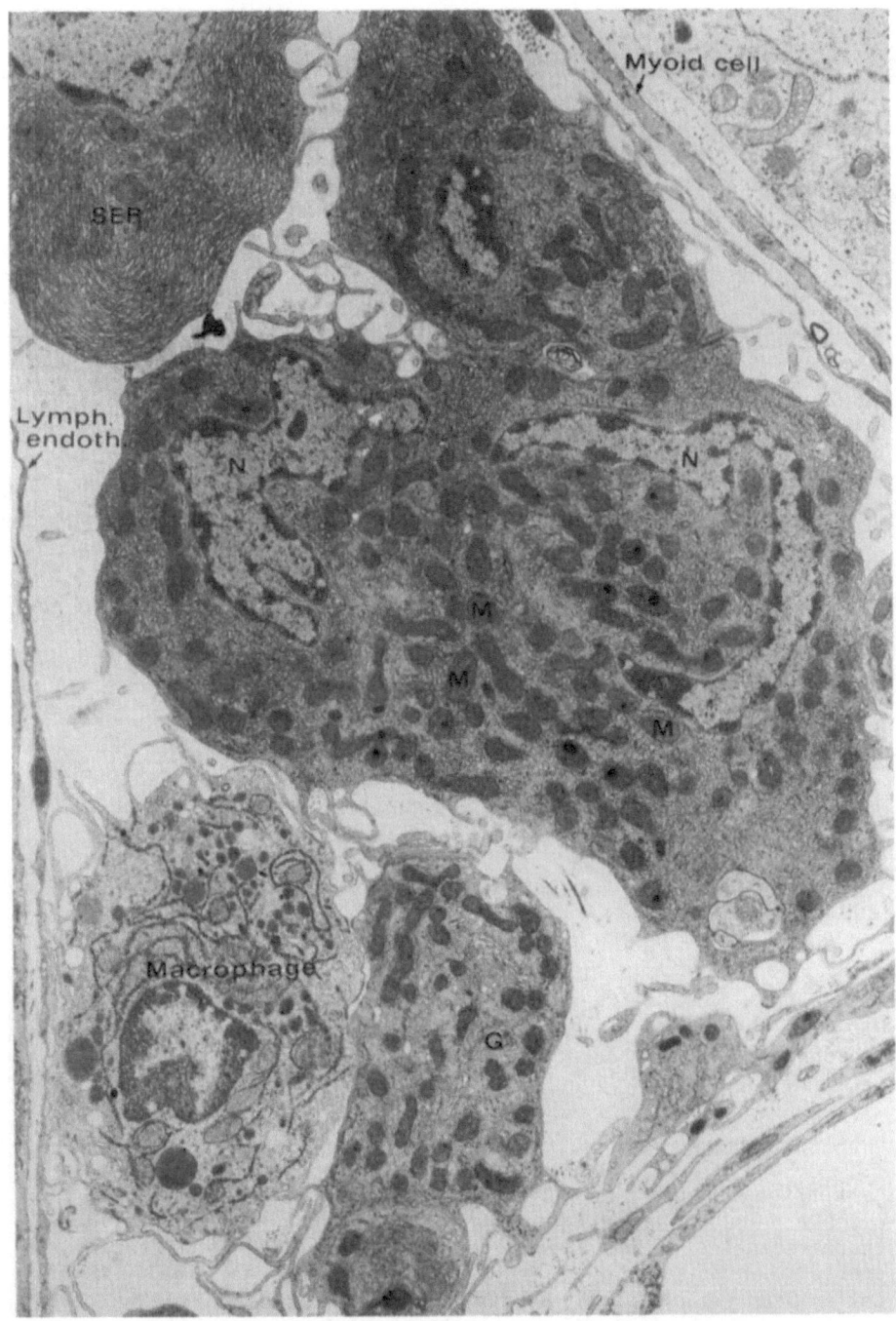

FIG. 2. Leydig cells from normal rats show nucleolus (NCL), smooth
(SER) and rough (RER) endoplasmic reticulum, mitochondria (M) with
lamellar and tubular cristae and Golgi complex (G).

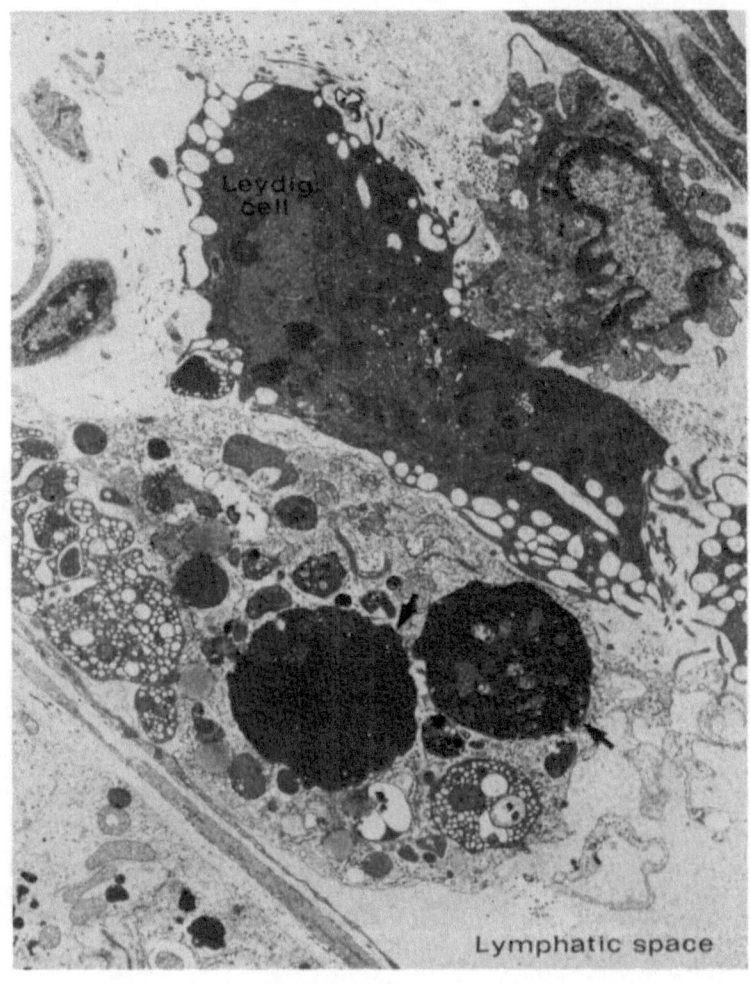

FIG. 3. Degenerating Leydig cells (arrows) are engulfed by macrophages (M), 24 hours after EDS.

As mentioned earlier, in the rat, the two generations of Leydig cells are not entirely separated in time although the cytological characteristics of the Leydig cells are sufficiently different to enable identification of foetal Leydig cells and adult Leydig cells (3). The foetal Leydig cell is characterized by large numbers of lipid inclusions and mitochondria with tubular cristae whereas the lipid content in adult Leydig cell is sparse and the mitochondrial cristae are lamellar.

14

The destruction of Leydig cells by EDS has provided a tool to study the origin of Leydig cells in the rat. Several investigators have documented that following a single administration of EDS, Leydig cells undergo degeneration to disappear from the testis after 72 hours, the degenerate cells being phagocytosed by the macrophages (Figure 1-3), (11-14). Our own studies utilised morphometry and hCG binding as techniques to determine that the destruction of Leydig cells was complete. (Figure 4). The elimination of Leydig cells was accompanied by a fall in serum testosterone levels (Figure 5) subsequently returned to normal from day 21 after EDS (11,14,15). The return of testosterone levels to normal was accompanied by a regeneration of Leydig cells as shown by our morphometric data (Figure 4).

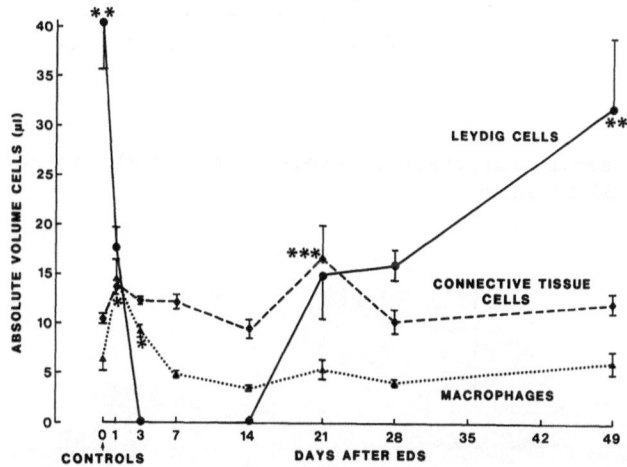

FIG. 4. The morphometric data shows the absolute volumes per testis of the intertubular cellular components at varying times after EDS.

Our own cytological studies have suggested that the new Leydig cells arise in a multifocal manner suggesting an origin by differentiation from a multiplicity of sources such as fibroblasts, pericytes and lymphatic endothelial cells (Figure 1 and 6) (14).

It is of interest that during regeneration of Leydig cells following EDS, the precursors differentiate into Leydig cells of the foetal type (Figure 7) large numbers of lipid inclusions and mitochondria with tubular cristae (13,14). Subsequently, 35-49 days after EDS, these foetal Leydig cells change their cytological characteristics to those adult Leydig cells. (Figure 8).

15

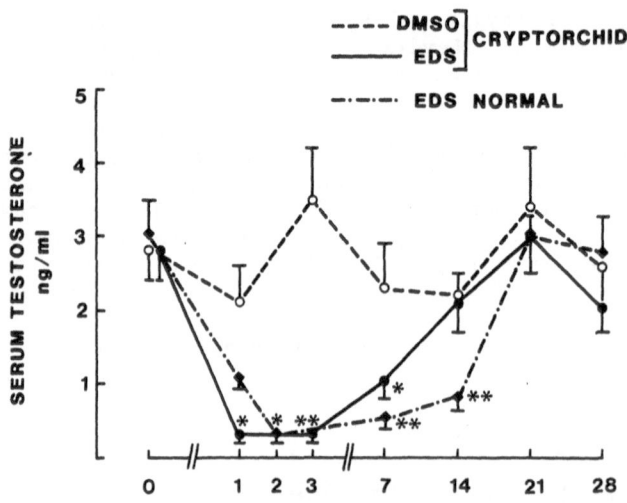

FIG. 5. The serum testosterone levels from normal and cryptorchid rats given EDS is shown.

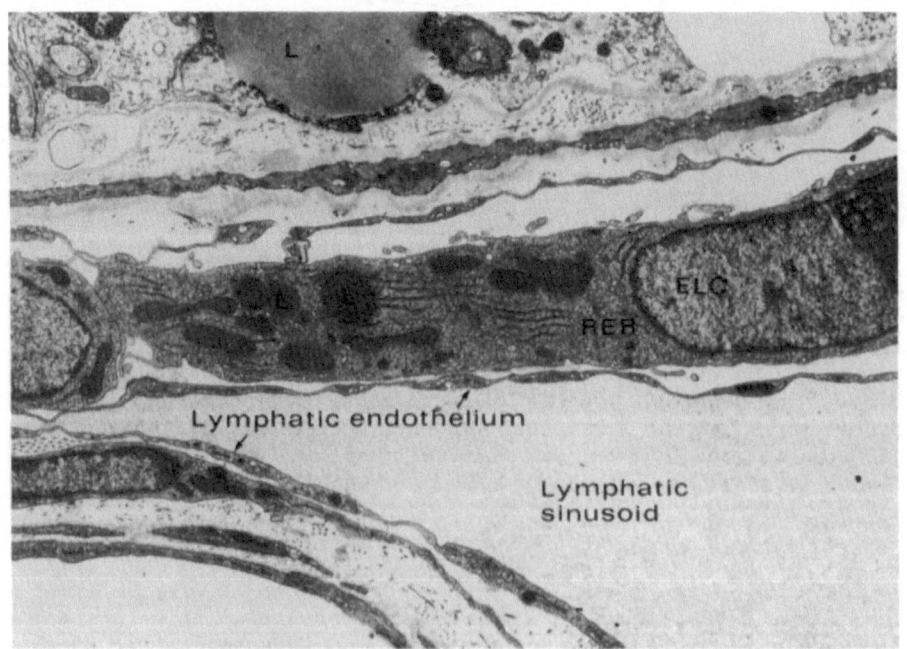

FIG. 6. The development of new Leydig cells 21 days post EDS occurs by differentiation from elongated cells (ELC). Note lipid (L), smooth endoplasmic and rough (RER) endoplasmic reticulum.

16

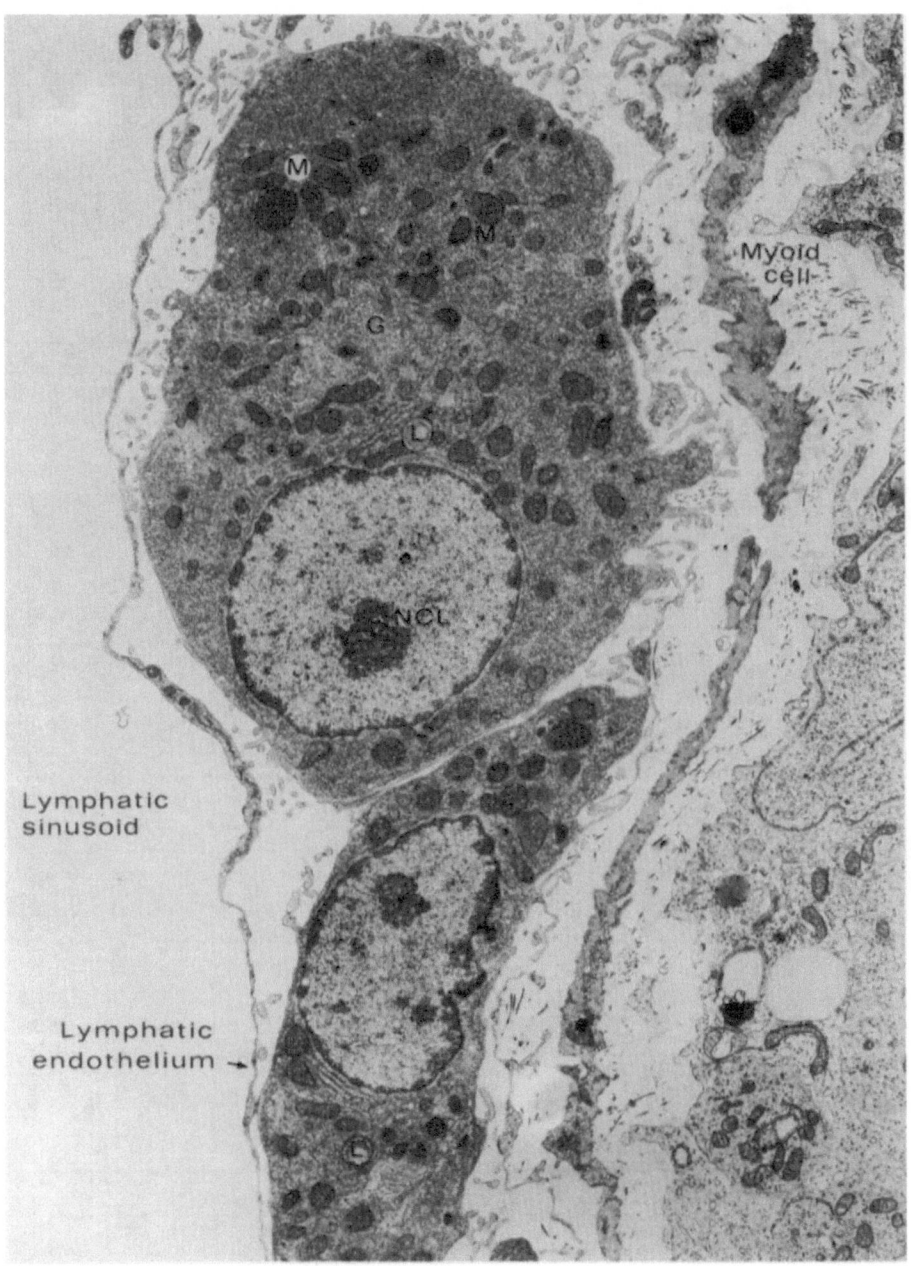

Fig. 7
Leydig cells 28–35 days post EDS show the characteristics of the
foetal type. Note lipid (L) smooth endoplasmic reticulum,
mitochondria with tubular cristae, nucleolus (NCL) and the Golgi
complex (G).

17

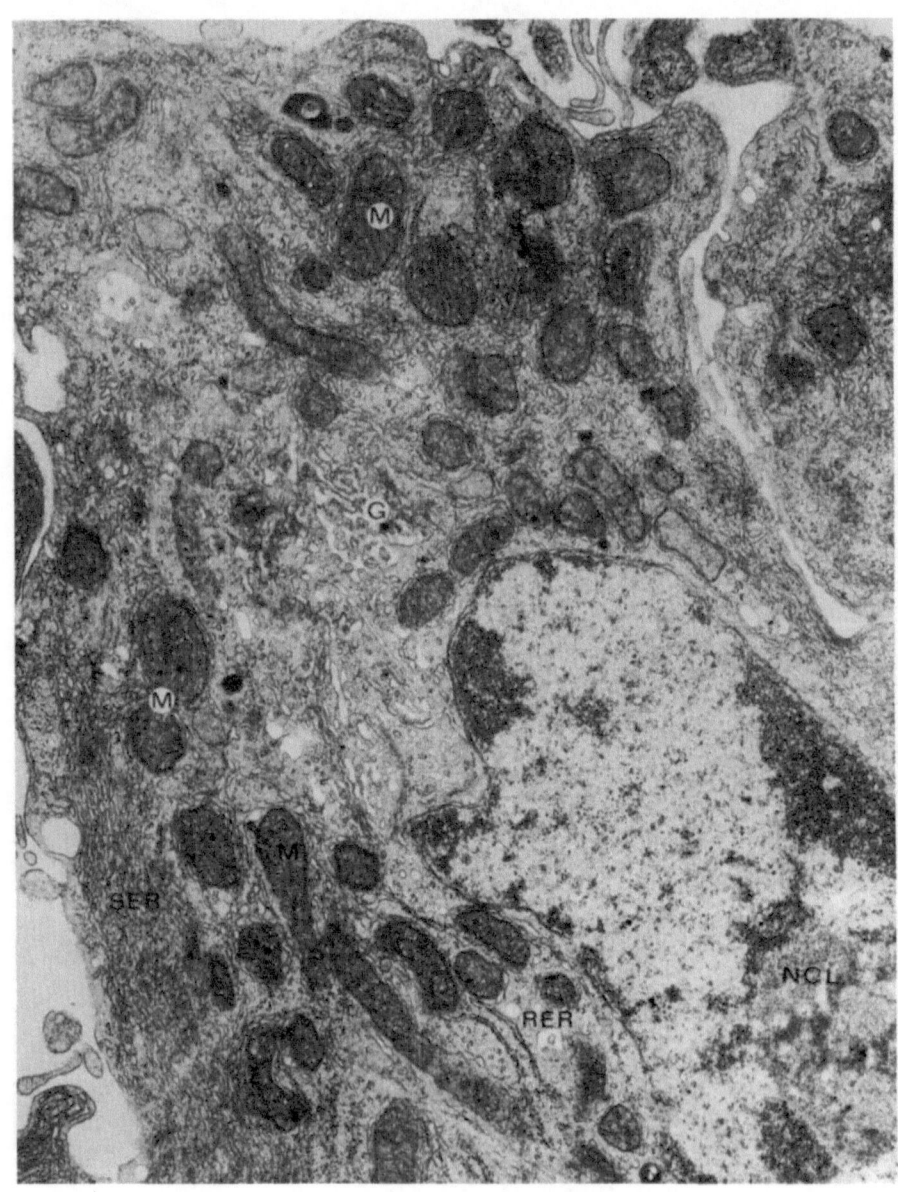

FIG. 8. After 35–49 days post-EDS the foetal Leydig cells change to the adult type. Note nucleus (N), mitochondria with lamellar and tubular-cristae (M), smooth endoplasmic reticulum (SER) and Golgi complex (G).

The regeneration of Leydig cells as shown cytologically, is supported by the results of ^{125}I-hCG binding which also reappears 21 days post EDS It is also of interest that serum testosterone levels return to normal before the absolute testicular volume of Leydig cells is restored to normal levels, suggesting that the elongated 'connective tissue precursors of Leydig cells are contributing to testosterone synthesis, a view in keeping with recent studies by Chemes and colleagues (16). Alternatively the testosterone output per "foetal type" Leydig cell must exceed that of the typical normal adult Leydig cell.

These studies with EDS indicate that cells within the intertubular tissue of the adult testis retain the ability to differentiate into Leydig cells when given the appropriate stimulus. The nature of this stimulus remains unclear but the recent study by Molenaar and colleagues (17) indicates that LH is a requirement for the regeneration of Leydig cells after EDS. This view is in keepng with data from Christensen and Peacock (18) who showed increased numbers of Leydig cells after chronic treatment of rats with hCG.

The results of these studies also indicate that the foetal type Leydig cells can differentiate into the adult variety, supporting the view that a similar process may occur in the transformation of the foetal generation to the adult generation of Leydig cells during sexual maturation in the rat, a process that occurs between days 20-25 of postnatal life (19).

Local control of Leydig cells by the Seminiferous tubules.
The concept that the seminiferous tubules could influence the Leydig cells was proposed by Aoki and Fawcett (20). Their morphological evidence was supported by extensive data morphological and functional data from our laboratories which showed that damage to the seminiferous tubule resulted in Leydig cell hypertrophy and hyperresponsivity of testosterone secretion to hCG stimulation (21-23) Further support for this concept arose from the studies of Bergh (24) who showed that the size of Leydig cells in the normal testis was influenced by the stage of the seminiferous cycle in the surrounding tubules. Maximum Leydig cell size was noted at stages VII-VIII of the rat seminiferous cycle.

Two possibilities exist, in one the seminiferous tubules are postulated to secrete an inhibitory factor acting on Leydig cells and following seminferous tubule damage this inhibition is removed. Alternatively, the seminiferous tubules may produce a factor capable of stimulating Leydig cells and following seminiferous tubule damage the secretion of this factor is increased. Over the past two years, significant support has emerged from the latter possibility. Sharpe & Cooper (25) demonstrated that testicular interstitial fluid from the rat could stimulate Leydig cell testosterone production and provided evidence that this substance was not LH since immunoneutralisaton with an antiserum to LH did not remove this activity. Similar data is available from our own laboratory supporting this concept

19

(Figure 9) and our unpublished data indicates that this factor has an apparent molecular weight of 30-35000 on gel filtration. Furthermore our data suggest that the action of this factor is early in steroidogenesis since it stimulates steroidogenesis prior to pregnenolone (26). Sharpe (27) has proposed that this factor is secreted in response to a lowering of intratesticular testosterone concentrations and is an attempt to conserve seminiferous tubule testosterone levels.

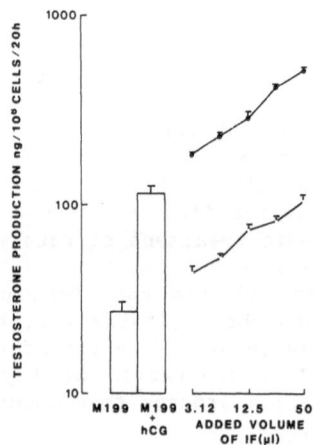

FIG. 9. The changes of testosterone secretion from Percoll-purified Leydig cells over 20 hours is shown in the presence or absence of a maximally stimulating dose of hCG. Note the further stimulation with interstitial fluid (IF) in the absence (o-o) or presence (•-•) of hCG.

We have used EDS to explore aspects of this local control mechanism. Our studies (28) and those of Molenaar and colleagues (17) indicated that regeneration of Leydig cells in testes following·spermatogenic damage was more rapid than that in the normal testis. Whereas Leydig cells returned in normal rats 21 days post EDS, in cryptorchid rats, Leydig cells reappeared after 14 days (Figure 5,10,11). These results suggest that the intertubular tissue in the cryptorchid rats was under the influence of a factor which was capable of causing a more rapid regeneration of Leydig cells following EDS. Similar experiments have also been performed following unilateral cryptorchidism and demonstrate that the return of Leydig cells is more rapid in the abdominal testis than in the scrotal testis (29) This further supports the view that a local factor is involved since circulating LH levels in these rats were indistinguishable from normal.

At this stage of our knowledge it is unclear whether the non-gonadotrophic factor in interstitial fluid capable of stimulating testosterone and the agent which causes a more rapid regeneration of Leydig cells following EDS treatment of cryptorchid rats are

20

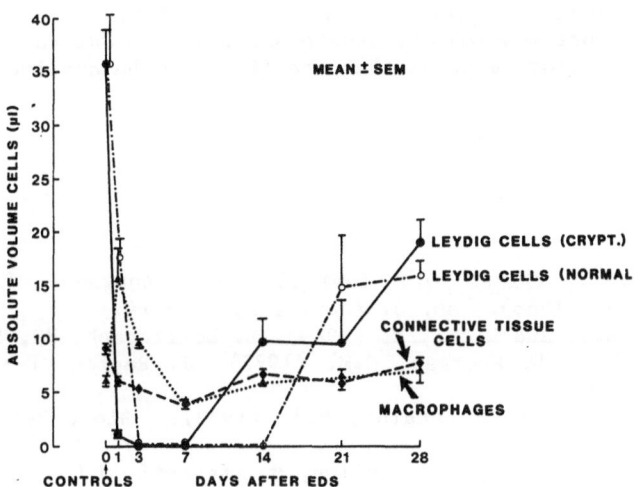

FIG. 10. The absolute volume per testis of cells from the intertubular tissue at varying times after EDS is shown from normal and cryptorchid rats.

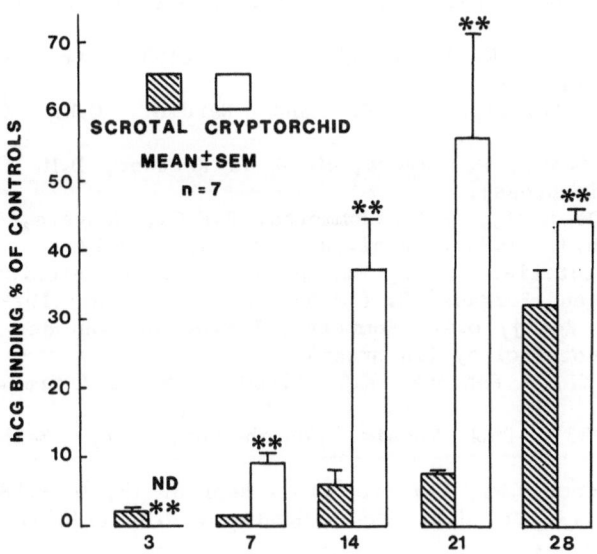

FIG. 11. The binding of ^{125}I-hCG to testicular homogenates from the testes (Scrotal and cryptorchid) of unilateral cryptorchid rats is shown at varying times after EDS.

related. Further work is necessary to characterize these factors though their purification may create difficulties due to the small volumes of starting material that are likely to be available.

REFERENCES

Mancini, R.E., Vilar, O., Lavieri J.C., Andrada, J.A. and Heinrich, J.J. (1963). Am. J. Anat. 112, 203-214.

Pelliniemi, L.J. and Niemi, M. (1969). Z. Zellforsch, 99, 507-522.

Lording, D.W. & de Kretser, D.M. (1972). J. Reprod. Fert. 29 : 261-269.

van Straaten, H.W.M. & Wensing, C.J. (1978). Biol. Reprod. 18, 86-93.

Mancini, R.E. Rosemberg, E., Cullen, M., Lavieri, J.C., Vilar, O., Gondos, B., Renston, R.H., & Goldstein, D.A. (1976). Am. J. Anat. 145, 167-182.

Chemes, H.E., Rivarola, M.A. & Bergada, C. (1976). J. Reprod. Fact, 46, 279-282.

Kerr, J.B., Robertson, D.M., and de Kretser, D.M. (1985). Endocrinology, 116 : 1030-1043.

Clegg, E.J. & McMillan, E.W. (1965). J. Endocr. 31, 299-300.

Prince, F.P., (1984). Anat. Rec., 209 , 165-176.

Morris, I.D. and McCluckie, J.A. (1979). J. Steroid Biochem. 10, 467-469.

Kerr, J.B., Donachie, K. & Rommerts, F.F.G. (1985). Cell & Tiss. Research, 242 : 145-156.

Morris, I.D., Phillips, D.M. and Bardin, C.W. (1986). Endocrinology.

Jackson, A., O'Leary, P., Ayers, M. & de Kretser, D.M. (1986). Biol. Reprod. (in press).

Molenaar, R., De Rooij, D.G., Rommerts, F.F.G., Reuvers, P.J. & van der Molen, H.J. (1985). Biol. Reprod. 33, 1213-1222.

Chemes, H.E., Gottlieb, S.E., Pasqualini, T., Domenichini, E., Rivarola, M.A., and Bergada, C. (1985). J. Androl, 6 : 102-112.

Molenaar, R. De Rooij, D.G., Rommerts, F.F.G. and van der Molen, H.J. (1986). Endocrinology (in press).

Christensen, A.K., & Peacock, K.C. (1980). Biol. Reprod. 22 : 383-391

Mendis, C. (1985). Ph.D Thesis. Monash University, Melbourne, Australia.

Aoki, A and Fawcett D.W., (1978). Biol. Reprod. 19, 144-158.

Risbridger, G.P., Kerr, J.B. and de Kretser, D.M. (1981). Biol. Reprod. 24, 534-540.

Risbridger, G.P., Kerr, J.B., Peake, R.A. & de Kretser, D.M. (1981). Endocrinol. 109, 1234-1241.

de Kretser, D.M. (1982). Int. J. Androl. Supp. 5, 11-17.

Bergh, A. (1982). Int. J. Androl. 5, 325-330.

Sharpe, R.M. (1986). Clinics Endocrinol & Metab. 15, 185-207.

Risbridger, G.P. Jenkin, G.J. & de Kretser, D.M. (1986). J. Reprod. Fert. (In press).

22

Sharpe, R.M. (1986). Clinics Endocrinol & Metab. 15, 185-207.
O'Leary, P.O., Jackson, A.E. and de Kretser, D.M. (1986b). Mol.
Cell Endocrinol. (in press).
O'Leary, P.O., Jackson, A.E. and de Kretser, D.M. (1986b). Mol.
Cell Endocrinol. (In press).

Carcinoma in situ of the testis: a review

N.E. SKAKKEBAEK[1,2], J.G. BERTHELSEN [1,3], A. GI-
WERCMAN[1], J. MÜLLER[1,4]

[1]Laboratory of Reproductive Biology, Rigshospitalet, Copenhagen;
[2]University Department of Paediatrics, Hvidovre Hospital, Copenhagen; [3]University Department of Obstetrics and Gynaecology, Herlev
Hospital, Copenhagen; [4]University Department of Paediatrics, Rigshospitalet, Copenhagen, Denmark

Introduction

Carcinoma in situ (CIS) of the testis was first recognised
as a preinvasive lesion in 1972 as a result of studies of
testicular biopsies in infertile Danish men (1,2). Several
reports from Europe, USA and Asia have confirmed that CIS
cells are premalignant germ cells with a high potential of
invasive growth (3-6). Approximately 50% of infertile men
with CIS have developed invasive germ cell tumours within
5 years (7,8).

Gross pathology: A testis harbouring CIS changes is often
atrophic. In typical cases the size of the testis is approximately 8-12 ml (normal range 15-35). However, in rare
cases CIS can be detected in testes with a size of more
than 20 ml. Except for the small size the gross pathology
is unremarkable.

Histology: The common fluids used for fixation of testicular biopsies (Stieve's, Bouin's or Cleland`s fluid) are
adequate for studies of CIS. However, it should be
stressed that the germinative epithelium is very poorly
preserved after fixation with formaldehyde. For conventional
histology no special staining is necessary to demonstrate
the characteristic CIS morphology (Figs.1a & 1b). However,
the diagnosis may be facilitated by immunohistochemical
staining of placental-like alkaline phosphatase (PlAP).
This enzyme is present in many CIS cells whereas PlAP
never has been demonstrated in cells of normal spermatogenesis (9-12) (Fig. 2). Cleland's fluid is not suitable
for demonstration of PlAP. Therefore, we use Stieve's
fixative for routine purposes because it provides good
preservation of nuclear morphology as well as preservation
of PlAP.

   In biopsies which have been appropriately fixed, the
CIS cells can easily be identified inside the seminiferous

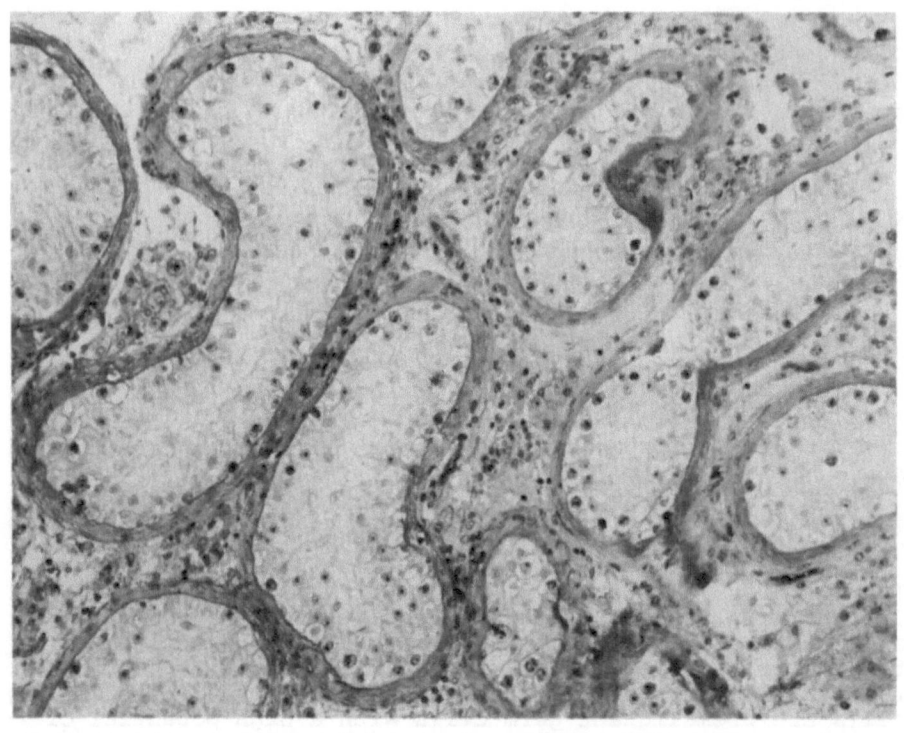

Fig. 1a. Section of testicular specimen showing carcinoma-in-situ in all tubules. Cleland's fixative, Iron-haematoxylin stain. x 170.

tubules (Fig. 1a). Tubules with CIS cells are scattered throughout the whole testis and comprise from a few to 100% of the tubules in the involved testis (13,14). In typical cases, CIS cells are located in a single row between normal appearing Sertoli cells and a thickened tubular membrane (Fig. 1b). The cytoplasm appears vacuolated owing to the extraction of glycogen accumulations. The CIS cells are larger than normal spermatogonia. The nuclear diameter is significantly higher (median 9.7 µm, range 9.2-10.5 µm) than that of spermatogonia (median 6.5 µm, range 5.7-7.1 µm) (15). The chromatin pattern of the nuclei is coarse and several nucleoli are present. Densitometric DNA-measurements reveal an aneuploid distribution pattern, similar to that of germ cell tumours (16-18).

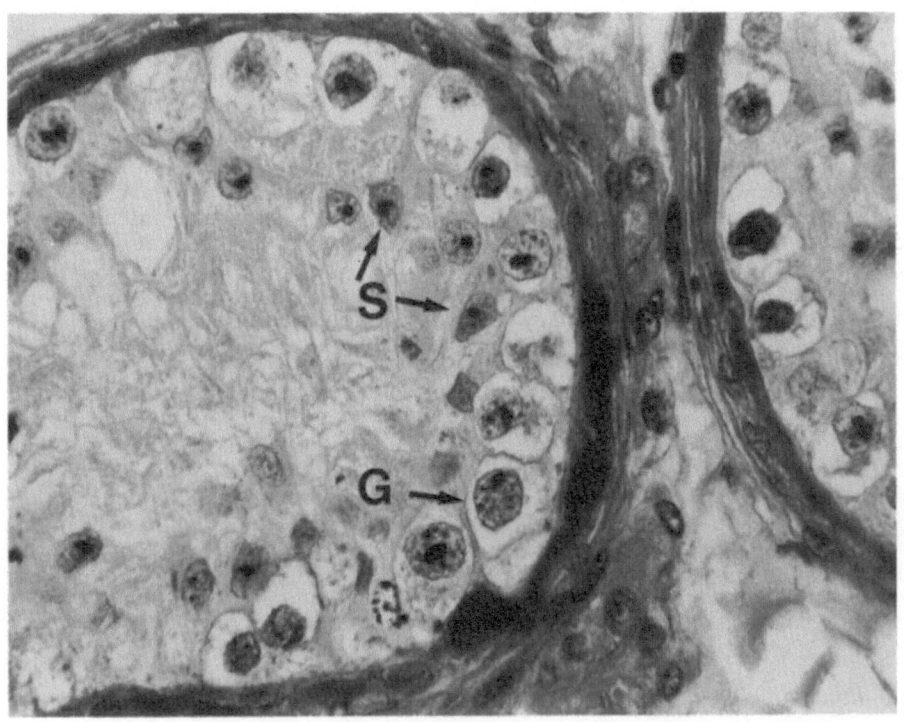

Fig. 1b. Same as 1a. Higher magnification. Note the
carcinoma-in-situ germ cells (G) and the Sertoli
cells (S). Note also the thickening of the
tubular membrane. x680.
From International Journal of Andrology, 1987,
10, in press, with permission.

Occasionally, CIS germ cells and normal germ cells are
found within the same tubule, representing a zone of transition between CIS epithelium and normal seminiferous
epithelium (14). Usually the morphology of the Sertoli
cells and the Leydig cells is normal. Sometimes foci of
lymphocytes are present in the interstitial tissue. Such
foci can be detected in testes without invasive growth,
although they are more common after invasive growth has
occurred (1,2,19). Invasive growth of CIS cells in the
tubular wall, the interstitial tissue or the rete testis
(Fig. 3) can be demonstrated even in the absence of a

macroscopic tumour (1,19-21).

CIS cells have also been demonstrated in prepubertal testes (22-25).

<u>Incidence</u>
The incidence of CIS in the general population is not known. However, on-going studies suggest that the incidence is much lower than in groups of individuals at increased risk of developing testicular germ cell tumours (A. Giwercman et al., unpublished data). These groups include men with a history of cryptorchidism (26-28),

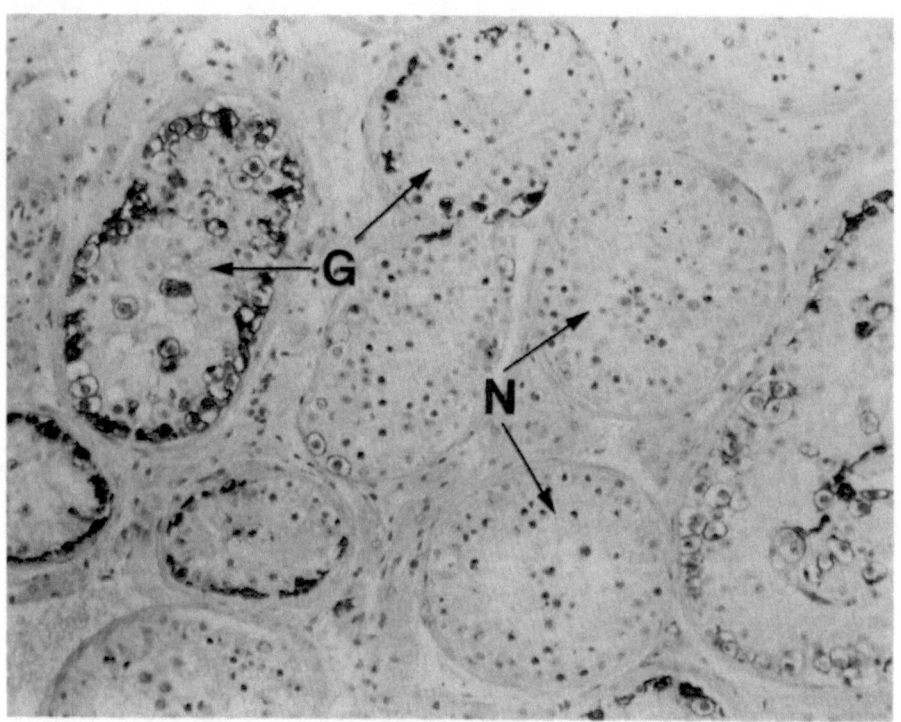

Fig. 2.   Section of testicular biopsy showing tubules with carcinoma in situ (G) and tubules with normal germ cells (N). Note the tubule in the middle of the Figure containing both normal and CIS germ cells. Immunoperoxidase demonstration of Placental-like alkaline phosphatase (DAKO No. A268) in CIS cells. PAP-technique. Stieve's fixative. x135

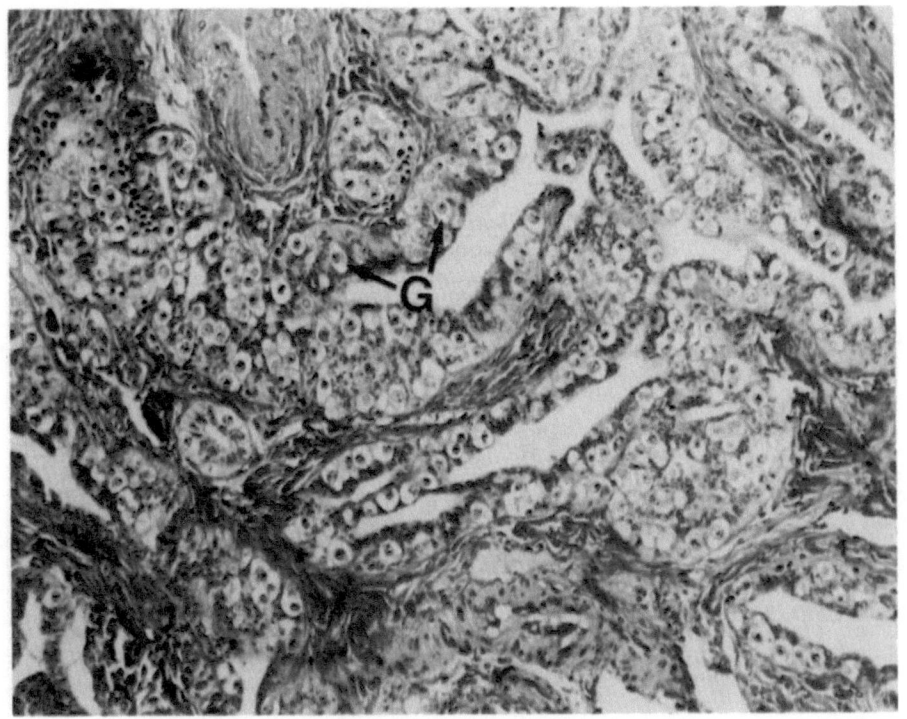

Fig. 3. Spread of carcinoma-in-situ to rete testis. Note
that the CIS germ cells (G) are located between
the cylindrical epithelial cells of rete and the
basal membrane. Cleland's fixative. x170.

patients with unilateral testis cancer (who have an
increased risk of CIS in the contralateral testis)
(29-30), infertile men (19,31,32) and individuals with
somato-sexual ambiguity (22,25,33).

Some recent studies indicate that CIS is present in
2-3% of young men with a history of treated or non-treated
maldescent of the testis (27,28)

In a prospective study of 600 consecutive patients
with unilateral testicular cancer we found CIS in the contralateral testis of approximately 5-6% (34).

Several groups have reported on CIS in testicular biopsies of infertile men. The true incidence of CIS among infertile men is not known. However, CIS was found in

0.4-1.1 % of testicular biopsies from selected infertile patients (19,31,32).

We have found a high incidence of CIS changes in pre-pubertal and pubertal children with androgen insensitivity syndrome (22) and 45,X/46,XY gonadal dysgenesis (25). However, the number of such patients studied so far is small.

Pathogenesis of testicular germ cell tumours.

CIS germ cells have been demonstrated in prepubertal testes, the youngest of these patients being only one month old (25). This finding has led to the hypothesis that the CIS cells are in fact malignant gonocytes. Hence, it has been suggested that testicular germ cell cancer may originate from germ cells of very early stages of embryonal life (11,35,36).

Many studies have shown that the CIS cells are precursors for both seminomas and nonseminomas (36) except spermatocytic seminomas (37).

The studies of patients with CIS have shown that, in untreated cases, the lesion will progress into invasive cancer in approximately 50% of cases within 5 years of observation period (7,8).

The association between CIS and manifest cancer has also been illustrated by the fact that seminiferous tubules with CIS can be detected in the testicular tissue adjacent to the majority of testicular tumours (38-41).

Early detection of testis cancer by screening for CIS

The treatment of testicular cancer often includes both radiotherapy and chemotherapy, which, in addition to the immediate side effects, may imply potential hazards to the future quality of life. Therefore, testicular neoplasia should preferably be diagnosed at the stage of CIS.

We now advocate routine screening of the contralateral testis for CIS in patients with germ cell tumours (29,30, 34). At present the role of screening in other risk groups such as patients with maldescent and infertility has not been established. However, on-going studies suggest that close follow-up including testis biopsy may be appropriate in young men with a history of maldescent (27,28).

Although a non-invasive method would be desirable, testicular biopsy is presently the only method of screening for CIS. A testicular biopsy is usually representative of the whole testis (13,14), and the risk of developing a testicular tumour in a testis without CIS in the biopsy seems to be extremely low (<1 in 1500).

Ultrasound and nuclear magnetic resonance may possibly in the future turn out to be an aid in screening for CIS of the testis (42,43).

Management

We recommend orchidectomy for unilateral CIS. If the

patient has only one testis or the CIS lesion has been found bilaterally, localised irradiation is the treatment of choice (34,44). We have used a fractionated radiation dose of 20 Gy. By this treatment CIS cells seem to be eradicated without any significant disturbance of Leydig cell function (34,44). A firm policy for the management of prepubertal individuals with CIS has not yet been established (34).

Concluding remarks ·
There seems to be little doubt that the incidence of CIS in the contralateral testis of patients with unilateral malignancy is approximately 5-6%. We believe that this incidence is high enough to warrant routine testicular biopsy at the time of orchidectomy for the primary malignancy. However, more research is needed to determine the incidence of CIS and define the role of screening for CIS in patients with undescended testes, infertility and somatosexual ambiguity.

References

1. Skakkebæk NE. Abnormal morphology of germ cells in two infertile men. Acta pathol microbiol scand [A] 1972: 80; 374-78.
2. Skakkebæk NE. Possible carcinoma-in-situ of the testis. Lancet 1972: ii; 516-17.
3. West AB, Butler MR, Fitzpatrick J, O'Brien A. Testicular tumors in subfertile men: Report of 4 cases with implications for management of patients presenting with infertility. J Urol 1985: 133; 107-09
4. Waxman M. Malignant germ cell tumor in situ in a cryptorchid testis. Cancer 1976: 38; 1452-56.
5. Dorman S, Trainer TD, Lefke D, et al. Incipient germ cell tumor in a cryptorchid testis. Cancer 1979: 44; 1357-62.
6. Ishida H, Isurugi K, Niijima T, Matsumoto K, Nomura K, Hirose K. Carcinoma in situ of germ cells and subsequent development of an invasive seminoma in a hyperprolactinaemic man. Int J Androl 1983: 6; 229-34.
7. Skakkebæk NE, Berthelsen JG, Visfeldt J. Clinical aspects of carcinoma-in-situ. Int J Androl 1981: Suppl 4; 153-63.
8. Skakkebæk NE, Berthelsen JG, Müller J. Carcinoma-in-situ of the undescended testis. Urol Clin N Am 1982: 9; 377-85
9. Beckstead JH. Alkaline phosphatase histochemistry in human germ cell neoplasm. Am J Surg Pathol 1983: 7; 341-49.

10. Jacobsen GK, Nørgaard-Pedersen B. Placental alkaline phosphatase in testicular germ cell tumours and in carcinoma-in-situ of the testis. Acta pathol immunol scand [A] 1984: 92; 323-29.
11. Skakkebæk NE, Berthelsen JG, Giwercman A, Müller J. Carcinoma-in-situ of the testis: possible origin from gonocytes and precursor of all types of germ cell tumours except spermatocytoma. Int J Androl 1987: 10; in press.
12. Hustin J, Collette J, Franchimont P. Immunohistochemical demonstration of placental alkaline phosphatase in various states of testicular development and in germ cell tumours. Int J Androl 1987: 10; in press.
13. Berthelsen JG, Skakkebæk NE. Value of testicular biopsy in diagnosing carcinoma in situ testis. Scand J Urol Nephrol 1981: 15; 165-68.
14. Berthelsen JG, Skakkebæk NE. Distribution of carcinoma-in-situ in testes from infertile men. Int J Androl 1981: Suppl 4; 172-84.
15. Müller J, Skakkebæk NE. Quantification of germ cells and seminiferous tubules by stereological examination of testicles from 50 boys who suffered from sudden death. Int J Androl 1983: 6; 143-56.
16. Müller J, Skakkebæk NE, Lundsteen C. Aneuploidy as a marker for carcinoma in situ of the testis. Acta path microbiol scand [A] 1981: 89; 67-68.
17. Müller J, Skakkebæk NE. Microspectrophotometric DNA measurements of carcinoma-in-situ germ cells of the testis. Int J Androl 1981: Suppl 4; 211-21.
18. Atkin NB. High chromosome numbers of seminomata and malignant teratomata of the testis: a review of data on 103 tumours. Br J Cancer 1973: 28; 275-79.
19. Skakkebæk NE. Carcinoma-in-situ of the testis: frequency and relationship to invasive germ cell tumours in infertile men. Histopathology 1978: 2; 157-70.
20. Nielsen H, Nielsen M, Skakkebæk NE. The fine structure of a possible carcinoma-in-situ in the seminiferous tubules in the testis of four infertile men. Acta pathol microbiol scand [A] 1974: 82; 235-48.
21. Schulze C, Holstein AF. On the histology of human seminoma. Cancer 1977: 39; 1090-1100.
22. Müller J, Skakkebæk NE. Testicular carcinoma in situ in children with the androgen insensitivity (testicular feminisation) syndrome. Br Med J 1984: 288; 1419-20.
23. Müller J. Morphometry and histology of gonads from 12 children and adolescents with the androgen insensitivity (testicular feminisation) syndrome. J Clin Endrocrinol Metab 1984: 59; 785-89.

24. Müller J, Skakkebæk NE, Nielsen O, et al. Cryptorchidism and testis cancer: atypical infantile germ cells followed by carcinoma in situ and invasive carcinoma in adult age. Cancer 1984: 54; 629-34.
25. Müller J, Skakkebæk NE, Ritzén M, et al. Carcinoma-in-situ testis in children with 45,X/46,XY gonadal dysgenesis. J Pediatr 1985: 106; 431-36.
26. Krabbe S, Skakkebæk NE, Berthelsen JG, et al. High incidence of undetected neoplasia in maldescended testes. Lancet 1979: i; 999-1000.
27. Pedersen KV, Boiesen P, Zetterlund CG. Experience of screening for carcinoma-in-situ of the testis among young men with surgically corrected maldescended testes. Int J Androl 1987: 10; in press.
28. Giwercman A, Berthelsen JG, Müller J, von der Maase H, Skakkebæk NE. Screening for carcinoma-in-situ of the testis. Int J Androl 1987: 10; in press.
29. Berthelsen JG, Skakkebæk NE, von der Maase H, et al. Screening for carcinoma in situ of the contralateral testis in patients with germinal testicular cancer. Br Med J 1982: 285; 1683-86.
30. Maase H von der, Berthelsen JG, Jacobsen GK, et al. Carcinoma in situ of the contralateral testis in patients with testicular germ cell cancer. Br Med J, in press.
31. Nüesch-Bachmann JH, Hedinger C. Atypische Spermatogonien als Präkanzerose. Schweiz Med Wochenschr 1977: 107; 795-801.
32. Pryor JP, Cameron KM, Chilton CP, Ford TF, Parkinson MC, Sinokrot J. Westwood CA. Carcinoma in situ in testicular biopsies from men presenting with infertility. Br J Urol 1983: 55; 780-84.
33. Nogales FF, Toro M, Ortega I, Fulwood HR. Bilateral incipient germ cell tumours of the testis in the incomplete testicular feminization syndrome. Histopathology 1981: 5; 511-15.
34. von der Maase H, Giwercman A, Müller J, Skakkebæk NE. Management of carcinoma-in-situ of the testis. Int J Androl 1987: 10; in press.
35. Skakkebæk NE, Berthelsen JG. Carcinoma in situ testis and development of different types of germ cell tumours. Adv Androl 1983: 7; 89-93.
36. Skakkebæk NE, Berthelsen JG, Müller J. Histopathology of human testicular tumours: carcinoma-in-situ germ cells and invasive growth of different types of germ cell tumours. In: Recent progress in cellular endocrinology of the testis. INSERM 1984: 123; 445-62.

37. Müller J, Skakkebæk NE, Parkinson MC: The
    spermatocytic seminoma: views on pathogenesis. Int
    J Androl 1987: 10; in press.
38. Mark GJ, Hedinger C. Changes in remaining tumor-free
    testicular tissue in cases of seminoma and
    teratoma. Virch Arch [A] 1965: 340; 84-92.
39. Skakkebæk NE. Atypical germ cell in the adjacent
    'normal tissue' of testicular tumours. Acta path
    microbiol scand [A] 1975: 83; 127-30.
40. Jacobsen GK, Henriksen OB, von der Maase H. Carcinoma
    in situ of testicular tissue adjacent to malignant
    germ-cell tumors: a study of 105 cases. Cancer
    1981: 47; 2660-62.
41. Coffin CM, Ewing S, Dehner LP. Frequency of
    intratubular germ cell neoplasia with invasive
    testicular germ cell tumors. Acta Pathol Lab Med
    1985: 109; 555-59.
42. Lenz S, Giwercman A, Skakkebæk NE, Bruun E,
    Frimodt-Møller C. Ultrasound in detection of early
    neoplasia of the testis. Int J Androl 1987: 10; in
    press.
43. Thomsen C, Jensen KE, Giwercman A, Kjær L, Henriksen
    O, Skakkebæk NE. Magnetic resonance: in vivo tissue
    characterization of the testis in patients with
    carcinoma-in-situ of the testis. Int J Androl 1987:
    10; in press.
44. von der Maase H, Giwercman A, Skakkebæk NE. Radiation
    treatment of carcinoma-in-situ of testis. Lancet
    1986: i; 624-25.

Fine needle aspiration biopsy of the testicles:
I. Cytological quantification

Z. PAPIĆ, G. KATONA, J. BARIŠIĆ, P. CVITKOVIĆ

Institute for Diabetes, Endocrinology and Metabolic Diseases "Vuk Vrhovac", Medical School University, Zagreb, Yugoslavia

Cytological aspiration biopsy, applied mainly for detection of testicular tumors, has been also used for the general insight into the subject's fertilizing ability(1). Further investigation in this field revealed the need for quantification (2), which was for the first time established in 1973.(3). Our diagnostic approach in cytologic evaluation of testicular smears in subjects with supposed infertility is based on these works and thorough investigation on specific forms of germinative and other cells(4). Cytological quantification revealed its diagnostic value both as in imprint smears and correlation with histological diagnosis(5,6), as well as correlated with other relevant andrological parameters(7).

Following sentences describe the procedure and methodology of the cytological quantification.

After cleaning of skin, testicles are punctured (one by one) with insulin needle and 10 cc syringe. Obtained material contains frequently also tubules, processed as histological sample (referred in part II.). Cellular, mostly liquid content of the needle is spread on one or more glass slides and stained after Pappenheim and Papanicolaou. Stained smears are quantified counting 500 consecutive cells, Sertoli cells included. Cytological finding contains numbers and indexes as follows:

1. SPERMATOGRAMME (Stg): The relation of particular developmental forms of spermatogenetic cells in 100 spermatogenetic cells.
2. MYTOTIC INDEX (MI): Number of spermatogenetic cells undergoing division in 100 spermatogenetic cells.
3. SPERM INDEX (SI): Number of sperms on 100 spermatogenetic cells.
4. SERTOLI CELLS INDEX (SEI): Number of Sertoli cells on 100 spermatogenetic cells.
5. SPERM - SERTOLI CELLS INDEX (SSEI): Number of sperms on 100 Sertoli cells.

The figure shows mutual relationship between listed indexes:

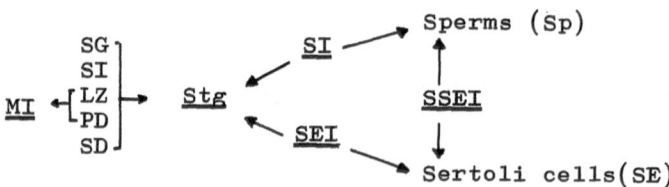

The quality of proposed diagnostic method is largely based on the reproducibility of identification of particular tubular cells(4,5). Cell forms identified (see figure) are: SG-spermatogonia, SI-primary spermatocyte, LZ-leptotene and zygotene forms in spermatocyte division, PD-pachytene, diachinesis and other forms of spermatocyte division and SD-spermatids. In that respect Stg will give the clinician – andrologist detailed insight in the sequence of development from spermatogonia till spermatids, indicating also disturbances in this sequence - different arrests. Sperm index (SI) displays content of sperms, relative to the spermatogenetic cells present presumably in the same tubular area. Increase of this index indicates the presence of desquamation and obturation in the testicle analysed. Low level of Sperm index(SI) is to be analyzed together with other indexes for better insight into underlying pathological condition. As number of Sertoli cells is constant per unit of tubular length indexes employing Sertoli cells are very helpful in defining cellular population of tubules. In that respect SEI indicates the quantity of germinal cells, i.e. the germinal potential, while SSEI gives the number of sperms i.e. the level of spermiogenesis.

The mutual relations of the indices may indicate some diagnostically relevant conditions. E.g., if both SI and SSEI are elevated, it indicates that the number of sperms is larger than local spermiogenesis could bear, so they are brought up by proximal part of the tubule and were retained due to an obturation. Both these indexes decreased indicates decreased spermiogenesis. Spermatogenic damage with a tendency towards tubular depopulation will be shown with elevated SEI, while arrests or tendencies towards an arrest will be specified with Stg. In cases where both SI and SEI are elevated, it indicates decreased spermatogenesis, because the number of spermatogenic cells is decreased in relation to the number of Sertoli cells. If SI is decreased while SEI is elevated this indicates decreased spermatogenesis and spermiogenesis. Whether it is caused by proportional damage to the germinative epithelium or by interruptions in sperm genesis will be shown by the Stg and MI. If SEI is elevated and SSEI decreased it means that the number of Sertoli cells is elevated. It could be the consequence of depopulation or early arrest.

Our experience with fine-needle aspiration biopsy of the testicles is rather ample, shown on the following table:

Year	1981	1982	1983	1984	1985	1986(Sept.)
Number of testicles	144	182	200	306	332	398

On the basis of these 1560 analysed cases and smears following indications for the puncture should be recommended:

1. Azoospermia
2. Oligozoospermia (below $20 \times 10^9/1$)
3. Evaluation of previous treatment effects.

It should be mentioned that in a considerable number of puncture cases the aspiration biopsy is repeated, which makes the total number of cases lower, but alowed more thorough analysis of particular cases and, maybe the most important, that this aspiration biopsy is virtually painless what makes patients prone to repeating of the sampling.

General advantages of testicular cytological quantification are relative and absolute numbers more detailed than any description, detailed insight into all tubular cell forms allowing minucuous analysis of e.g. effects of treatment. The possibility of simple repeating allows morphological control whenever required.

BIBLIOGRAPHY

1. Posner,C.,Cohn,J. Zur Diagnose und Behandlung der Azoospermie. Dtsch.med.Wschr.,1904:29;1062-87

2. Škrabalo,Z. Endokrinološki poremećaji muških spolnih organa u toku teških konzumptivnih oboljenja (Endocrinological Disturbances of Male Gonads During Consumptive Diseases). Ph.D.Thesis, Medical School University of Zagreb,1962:101-3

3. Črepinko,I.,Posinovec,J.,Škrabalo,Z. Cyto-histological Comparison in Sterile Human Testes.Acta med.iug., 1973:27;449-54.

4. Papić,Z. Morfološka studija oligospermija i azoospermija (Morphological Study of Oligospermia and Azoospermia). Ph.D.Thesis, Medical School University of Zagreb, 1984:96-100.

5. Papić,Z.,Katona,G. Testicular Cell Morphology:Its Quantification, Reproducibility and Relation to Routine Morphological Tests. Acta Cytologica (in print).

6. Papić,Z.,Katona,G.,Barišić,J.,Cvitković,P. Fine-needle Aspiration Biopsy of·the Testes in the Diagnosis of
Fertility with the Possibility of Histologic Evaluation
of the same Material. Citologia, 1985:7 (Suppl.);167-7.

7. Cvitković,P.,Gavella,M.,Papić,Z.,Singer,Z.,Škrabalo,Z.
Seminal Plasma Isoenzyme LDH-X Values in Extreme Oligo-
zoospermia.In:The Male Factor in Human Infertility Diagnosis and Treatment,eds. Thompson,W.,Harrison,R.F.,Bon-
nar,J. Lancaster,Boston,The Hague,Dordrecht,MTP Press
Limited,1983:271-74.

38

Fine needle aspiration biopsy of the testicles:
II. Histological evaluation of the same material

G. KATONA, Z. PAPIĆ, S. VLAHOVIĆ, P. CVITKOVIĆ

Institute for Diabetes, Endocrinology and Metabolic Diseases "Vuk Vrhovac", Medical School University, Zagreb, Yugoslavia

In the previous article it has already been mentioned that during the ejection of the needle content on the glass slide, tiny pieces of tissue can quite frequently be found. This is due to a specific structure of the testicular parenchyma, i.e. tubules. We have tried to collect them carefully with the same needle and to process them as in any small needle biopsy. The effect was positive and after some preparation we have introduced the technique into the morphological methods used for evaluation of human fertility and diagnosis of infertility. It has now been routinely used for five years and we would like hereby to present the method, on the basis of 820 biopsies obtained and analyzed so far. We would also like to present our method of numerical expression of the histological diagnosis.

The general rate of material obtained is 3/4, i.e. 0.75 of the cytological aspiration biopsies. Analyzing the cases we did not get the material for histology, it appears that their morphological changes were quite developed, probably in terms of general fibrosis, tubular shrinkage, etc. In such cases, only surgical biopsy could be advised for the collection of sufficient material. However, surgical biopsy is rather painful, while fine needle biopsy is not. It is rather unlikely that the patients would accept repeated surgical biopsy, while they come quite frequently to repeated needle biopsy, thus allowing us to analyze the effects of treatment, other changes in the tissue, etc. The material can also be used in cytogenetic studies (1). Concerning stainings, H&E and Mallory have been used routinely. To date, our results have been reported at endocrinological (2) and cytological (3) meetings, and recommended in a manual (4). A comparison with the cytological findings obtained on the same material revealed a good correlation, with the exception of cases with mild to medium depopu-

THE SCHEME OF EVALUATION OF TESTICULAR BIOPSY HISTOLOGICAL FINDING

TUBULES	INTERSTITIUM	CONTENT	DEVELOPMENT
1.Normal	1. Normal	1. Normal	1. Normal
2.Normal with fibrosis	2. Mild focal fibrosis	2. Desorganisation	2. Spermatogenetic damage
3. Narrow	3. Medium focal fibrosis	3. Desquamation	3. Partial late arrest
4. Narrow with thickening	4. Strong focal fibrosis	4. Desquamation and obturation	4. Late arrest
5. Narrow with strong thickening	5. Inflammation	5. Mild depopulation	5. Partial early arrest
6. Very narrow	6. Inflammation with focal fibrosis	6. Strong depopulation	6. Early arrest
7. Very narrow with thickening	7. Leydig cell hyperplasia	7. No sperms	7. No germinative cells
8. Very narrow with strong thickening		8. Sertoli cells only	
9. Total fibrosis		9. Sertoli cells with hyperplasia	

lation of germinative cells, which could be histologically seen only. As such, the method is very useful, practical and beneficial for patients.

Another method we would like to present, connected with the same topic and carried out on the same material, is the numerical expression of histological diagnosis. There are many good reasons to use the standardized numerical expression. First are the computer analysis and comparisons with hormonal, biochemical and cytological findings. The second is the standardization of diagnostic expressions, which is most exact if it is numerical. Histological diagnosis of testicular biopsy is at present mainly descriptive, containing at least four elements: tubular size and wall thickness, interstitial changes, and changes in tubular cellular content and in the developmental sequence of germinative cells. However, previously published numerical expressions (5,6) were of one digit only, allowing the maximum of 10 diagnoses. As it seemed somewhat limited for our purposes, we have introduced a four digit numerical expression, each digit covering one element of the diagnosis, with numbers starting from 1 = normal and growing so that the biggest number indicates the worst condition. It has been done in each group separately, so that each element is described independently.

The table presents the four elements: tubules, interstitium, cell content and cell development as well as the diagnosis within them. The same table is printed on the back of our finding forms.

The first two digits, description of the tubular wall and interstitial changes, mainly describe the case history. Obviously, it takes some time for the thickening of the tubular wall to develop or to produce focal fibrosis. At present, we cannot estimate either the duration of the disease or the biological importance of these particular changes. It is quite possible that the between-case differences are merely biological differences. Here a lot is to be understood.

The last two digits are better understood, more diagnosis exist here. Taken together, all the four digits represent a brief but detailed description of the biopsy. The scheme, of course, could also be used for surgical biopsies, but in that case a sort of a summary is to be prepared, and then the numerical expression represents a description of the summary. Nothing difficult!

BIBLIOGRAPHY

1. Singer,Z.,Cvitković,P.,Papić,Z.,Katona,G.,Škrabalo,Z. Karyotype 46,X,del(Y) (p ter - q 11:) in men with normal phenotype and azoospermia. Diab.Croat.,1984:13; 213-221.

2. Papić,Z.,Katona,G.,Črepinko,I.,Cvitković,P.,Škrabalo,Z.
Testis morphology in the diagnosis of male infertility.
Proceedings of the 3rd Yugoslav Congress on endocrino-
logy,1984:738-742.

3. Papić,Z.,Katona,G.,Barišić,J.,Cvitković,P. Fine-needle
aspiration biopsy of the testes in the diagnosis of
fertility with the possibility of histologic evaluation of the same material. Citologia, 1985:7 (Suppl.);
166-7.

4. Katona,G.,Papić,Z. The method of evaluation of small
histological testicular samples. Diab.Croat, 1979:
8 (Suppl. II);25-27.

5. Girgis,S.M.,Hafez,E.S.E. Evaluation of testicular
biopsy. In: E.S.E.Hafez: Techniques of human andrology.
North-Holland publishing company, Amsterdam-New York-
-Oxford,1977:83-112.

6. Posinovec,J.,Škrabalo,Z. Morphological aspects of
sterility in men.Acta med.iug.,1974:28;321-328.

A quantitative histological investigation of the Sertoli cells during human development

D. CORTES[1], J. MÜLLER[1], N.E. SKAKKEBAEK[1,2]

[1]*Laboratory of Reproductive Biology, Rigshospitalet, Copenhagen;*
[2]*University Department of Paediatrics, Hvidovre Hospital, Copenhagen, Denmark*

Only a few studies have dealt with quantitative changes in the population of Sertoli cells during human development (1,2,3). The results of these studies are conflicting. According to one established theory the number of Sertoli cells remains constant during postnatal development while only a maturation of the cells takes place during puberty (2). Using stereological methods for quantification (4,5, 6), we have carried out a study of the total number of Sertoli cells and the size of their nuclei on testicular tissue obtained from 5 foetuses who were stillborn, and 31 individuals between 3 months and 40 years of age, who had suffered from sudden, unexpected death. The numerical density of Sertoli cells, the total number of Sertoli cells per individual, and the mean nuclear volume of the Sertoli cells were determined by point- and profile counting of sections of 0.5 µm. The total number of Sertoli cells per individual rose significantly from a median of 260 mill (range 130-520 mill) during the late foetal period to 1500 mill (860-2900 mill) in individuals 3 m - 10 yr of age. A further increase was found during puberty as the number of Sertoli cells in adults (>18 yr) was 3700 mill (2500-5600 mill). The numerical density of Sertoli cells decreased from a median of 1200 mill/cm^3 during childhood (3 m - 10 yr) to 140 mill/cm^3 (110-260 mill/cm^3) in adults (>18 yr). The mean nuclear volume of Sertoli cells increased from a median of 100 µm^3 (60-170 µm^3) during the 3 m - 10 yr period to 210 µm^3 (170-260 µm^3) in adults (>18 yr). These results indicate that significant quantitative and qualitative changes in the population of Sertoli cells take place after birth.

References

1) Hadžiselimović F & Seguchi H. Ultramikroskopische untersuchungen an Tubulus Seminiferous bei Kindern von der Geburt bis zu Pubertet. II. Entwicklung und Morphologie der Sertolizellen. Verh Ana Ges 1974: 68; 140-161.

2)   Nistal M, Abaurrea MA & Paniagua R. Morphological and histometric study on the human Sertoli cell
from birth to the onset of puberty. J Ana 1982: 14;
351-363.

3)   Läckgren G & Plöen L. The influence of human chorionic gonadotropin (hCG) on the morphology of the prepubertal human undescended testis. Int J Androl 1984: 7;
39-52.

4)   Weibel ER. Stereological Methods. Vol 1.
Practical Methods for Biological Morphometry. Academic
Press, London, 1979.

5)   Weibel ER. Stereological Methods. Vol 2.
Theoretical Foundations. Academic Press, London, 1980.

6)   Müller J & Skakkebæk NE. Quantification of
germ cells and seminiferous tubulus by stereological
examination of testicles from 50 boys who suffered
from sudden death. Int J Androl 1983: 6; 143-156.

Hypogonadotropic status and testicular morphology in nonhuman primates (M. mulatta and M. fascicularis)

G.F. WEINBAUER, M. RESPONDEK[1], H. THEMANN[1], E. NIESCHLAG

Max Planck Clinical Research Unit for Reproductive Medicine and Department of Experimental Endocrinology, University Women's Hospital, Münster; [1]Department of Medical Cytobiology, University of Münster, Münster, FRG

The testicular morphology of non-human primates with hypogonadotropic hypogonadism experimentally induced by GnRH analog treatment, hypophysectomy (HY) or pituitary stalk-sectioning (PSS) was studied using light and electron microscopy. The hypogonadotropic status was verified for all treatments by measurement of bioactive LH and testosterone in serum (1,2,3,4).

Histological Examination

The testicular tissue was obtained either through biopsy or at sacrifice from cynomolgus monkeys (M. fascicularis) treated with GnRH antagonist or after HY, or from rhesus monkeys (M. mulatta) under GnRH agonist treatment or following PSS. The testicular specimens were fixed in glutaraldehyde, postfixed in osmium tetroxyde, dehydrated in ethanol and embedded in epoxy resin. Semithin sections (1 um) were stained with toluidine-blue and ultrathin sections with uranyl acetate and lead citrate. The diameter of 20 seminiferous tubules per specimen was measured using a semi-automatic image analysis system.

GnRH Antagonist versus Hypophysectomy

The reduction in tubular diameters after HY was much more pronounced than after GnRH antagonist treatment (Table I). Remarkably, the diameters were virtually iden tical 76 and 180 days after hypophysectomy.

Table I. Seminiferous tubule diameter and spermatogenesis of cynomolgus and rhesus monkeys with hypogonadotropic hypogonadism

Treatment	Duration (days)	Tubular Diameter (um)	Spermatogenesis	Species
GnRH antagonist	63		disruption at level of spermatocytes; occasionally sperma-tids	cynomolgus
# 3919		124 ± 19		
# 3911		144 ± 34		
# C		148 ± 24		
# 4		161 ± 23		
Hypophysectomy			disruption at level of spermatogonia; no spermatocytes and spermatids	cynomolgus
# 4302	76	79 ± 7		
# 3917	180	71 ± 13		
GnRH agonist	602		disruption at various cellular levels; spermatocytes and spermatids present	rhesus
# 910		105 ± 14		
# 3753		113 ± 27		
# 2899		137 ± 26		
# 2831		131 ± 14		
Stalk-section	488		disruption at level of spermatogonia; no spermatocytes and spermatids	rhesus
# 777		69 ± 12		
# 800		80 ± 12		
# 711		88 ± 19		

Tubular diameter (mean ± SD) was determined in tubules per specimen.

Spermatogenic arrest in GnRH antagonist-treated monkeys occurred mainly at the level of spermatocytes. Occasionally spermatids were present which appeared to be degenerating. In contrast, following HY the tubular epithelium was comprised of spermatogonia and Sertoli cells alone. Sertoli cell morphology after HY was characterized by nearly round nuclei and reduced numbers of cytoplasmic organelles. The appearance of these cells resembled that of immature Sertoli cells. Such morphological changes did not occur following GnRH antagonist treatment.

GnRH Agonist versus Pituitary Stalk-Sectioning

Tubular diameters from PSS-rhesus monkeys were about 75 % of those measured after GnRH agonist treatment even though the treatment period was longer by about 15 weeks in the latter (Table I). As in cynomolgus monkeys, the spermatogenic disruption after PSS invariably occurred at the spermatogonial level while with the GnRH agonist haploid germ cells could still be seen. Morphological signs of immaturity were only seen in Sertoli cells from PSS-testis.

In testes from all treatment groups the distribution of lipid and secondary lysosomal granules within the seminiferous tubules showed typical alterations. As tubular regression advanced the granules were found in increasing amounts towards the lumen or were scattered throughout the entire tubule. In cynomolgus monkeys the lipid droplets prevailed while in the rhesus monkey large, confluent, lysosomal vesicles predominated.

A thickening of the tubular wall was found in all treatment groups but was more pronounced following HY and PSS than after GnRH analog treatment.

Conclusions

The observations demonstrate that in spite of apparently similar hypogonadotropic status and irrespective of macaque species, testicular regression is more advanced following HY and PSS than after treatment with GnRH analogs. It may be speculated that the discrepancies in testicular involution are due to the lack of pituitary hormones other than gonadotropins.

References

1. Weinbauer GF, Surmann FJ, Akhtar FB, Shah GV, Vickery BH, Nieschlag E. Reversible inhibition of testicular function by a gonadotropin hormone-releasing hormone antagonist in monkeys (M. fascicularis). Fertil. Steril. 1984:42;906-914.

2. Weinbauer GF, Akhtar FB, Respondek M, Nieschlag E. Testicular morphology in GnRH agonist-treated monkeys. J. Androl. 1985:6; P-37

3. Marshall GR, Jockenhövel F, Lüdecke DK, Nieschlag E. Maintenance of complete but quantitatively reduced spermatogenesis in hypophysectomized monkeys by testosterone. Acta Endocrinol. (in press).

4. Marshall GR, Wickings EJ, Lüdecke DK, Nieschlag E. Stimulation of spermatogenesis in stalk-sectioned rhesus monkeys by testosterone alone. J. Clin. Endocrinol. Metab. 1983:57;152-159.

Localization of transferrin and somatomedin-C receptors in human seminiferous epithelium

G.B. VANNELLI, T. BARNI[1], C. ORLANDO[1], A. NA-TALI[2], M. SERIO[1], G.C. BALBONI

Institute of Human Anatomy; [1]Andrology Unit, Department of Clinical Physiopathology; [2]Department of Urology, University of Florence, Florence, Italy

INTRODUCTION

Sertoli cells in culture make Transferrin and Somato-medin-like peptides (1,2).
Transferrin (TF) is a plasma protein which provides the transport of iron, an essential growth factor for cell proliferation. The specific TF-receptor, a glycoprotein localized on the cell membrane, ·is essential for the transport of TF into the cell. This receptor is present only in proliferating cells (3).
Somatomedin C (Insulin like growth factor I or IGF-1) is another important protein able to stimulate the cell proliferation and differentiation.
Receptors for both proteins were demonstrated in rat pachytene spermatocytes but not at the level of sperma-togonia, differentiated spermatids and spermatozoa (4,5).
The receptors have not been demonstrated in humans.

MATERIALS AND METHODS

PATIENTS. Testicular biopsies were obtained under epi-dural anaestesia from patients (n=8) investigated for couple infertility. In two patients we found normal spermatogenesis, in one patient the picture of Sertoli Cell only Syndrome and in the others different degre-es of germinal arrest.

REAGENTS. Anti-human TF polyclonal antiserum (T-6265) developed in goat was obtained from Sigma Chemical Compa-

ny (St. Louis.MO).
Murine monoclonal antibody B3/25 (a92121) to human TF
receptor was obtained from Hybritech Inc. (San Diego,CA).
Murine monoçlonal antibody to human Somatomedin C (SMC)
receptor aIR-3 was prepared by Dr. S.Jacobs (The Wellcome Research Laboratories, Research Triangle Park, NC
U.S.A).
Antigoat IgG peroxidase conjugate (CA-8150) antimouse
IgG peroxidase conjugate (A-2028), human TF (T-2252)
were obtained from Sigma Chemical Company (St.Louis,MO).
3,4,3¦4'tetra-aminodiphenylhydrochloride (Diamino-benzi-
dine) was obtained from BDH Chemical Ltd. (Poole,England).
Goat antibody to mouse IgG (GAR 15) and rabbit antibody
to goat IgG (GAR 15) linked to colloid gold particles
were obtained from Janssen Life Sciences Products
(Beerse, Belgium).

PROCEDURES. The testicular fragments were fixed in Bouin's
solution and embedded in paraffin. Sections of about 5
microns of thickness were used for staining. Counter
stain was not used.
Immunoperoxidase staining was performed as described
by Sternberger (6) according to the following protocol.
Dilutions of antisera were carried out with 0.01 mol/L
phosphate buffered saline, pH 7.4 (PBS). Paraffin sections
were deparaffinized immediately before use. After rehy-
dratation separate sections were rinsed in PBS and then
incubated for 24 hours at 24 °C with anti-TF antiserum
(1:400), anti-TF receptor antibody (1:1000) and aIR-3
(1:100) respectively.The sections rinsed with buffer
were then incubated with the specific IgG-peroxidase
conjugate (1:30) for 30 minutes at room temperature.
After a new rinsing with PBS the sections were washed
again with 0.05 mol/L Tris-HCL buffer, pH 7.6. The
peroxidase activity was demonstrated with a solution
of 0.05% of diaminobenzidin in 0.05 mol/L Tris-HCL
buffer containing 0.01% hydrogen peroxide.
Positive results were controlled by: 1) incubation
with a non-immune serum, 2) using peroxidase-conjugate
alone, 3) serial dilutions of specific immune sera.
For the electron microscopic observation some specimens
were fixed in Karnovsky and embedded in Epon. The thin
sections were collected on nickel grids, etched with
10% hydrogen peroxide for 10 min. and rinsed twice in
PBS (pH 7.4). The grids were then placed, tissue side down
on drops of antiserum to human TF or TF-receptor (1:400)

and stored at 4 °C for 48 hours in a humified chamber.
Following two washes, rabbit anti-goat IgG (for TF detection) or goat anti-mouse IgG (for TF receptor detection)
both labelled with colloidal gold (1:100) for 2 hours
at room temperature. The gold particles were 15 nm in
diameter. The grids were washed three times in distilled water and stained with 5% uranyl acetate for 15 min.

RESULTS

In the sections treated with anti-human TF antiserum the
cytoplasm of Sertoli cells appeared prominent as an intensely reactive flame-shaped expansion toward the tubular lumen (Fig.1). A less evident positivity may be detectable also in the cytoplasm of 1) large round cells
with nuclei containing coarse patches of chromatin and
interpreted as pachytene primary spermatocytes and 2)
smaller round cells with condensed nuclear chromatin,
secondary spermatocytes and early spermatids in nature
(Fig.1).
As concern the specimens prepared using the monoclonal anti-TF receptor antibody, a sharp positive reaction
was observed in the same round large and small cells
previously interpreted as spermatocytes I and II and

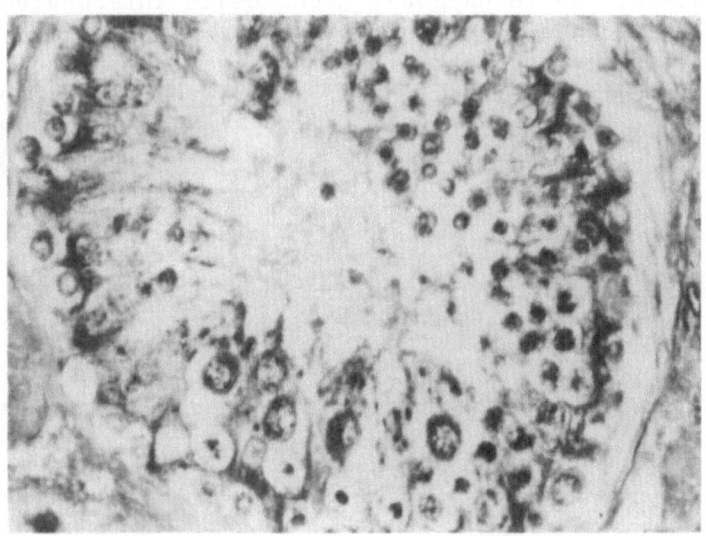

Fig.1. Immunostaining of TF (Immunoperoxidase Technique) in a testicular tubule of a subject with
normal spermatogenesis, using a polyclonal
antiserum anti-TF. The cytoplasm of Sertoli
Cells and round cells in the lumen are stained.

51

early spermatids (Fig.2).
All the controls for TF and TF receptors were always negative.
The studies performed using the electron microscopy demonstrated the presence of TF labelled colloidal gold in the cytoplasm of Sertoli cells (Fig.3).

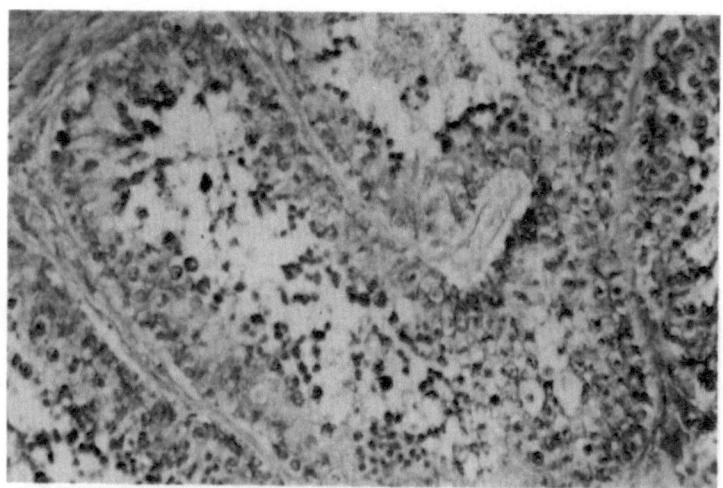

Fig.2. Section of a testicular tubule treated with monoclonal antibody anti-TF receptor (Immuno-peroxidase Technique). Round cells interpreted as secondary spermatocytes and early sperma-tids are stained.

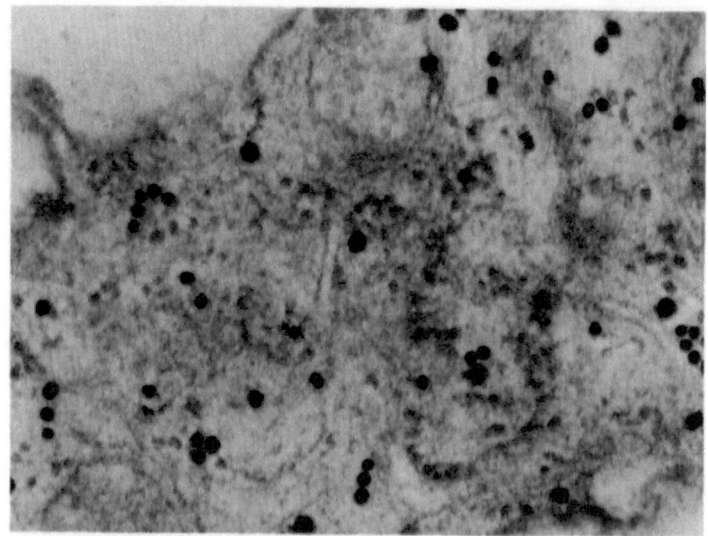

Fig.3. Cytoplasm of Sertoli cell of a normal human testis treated with TF polyclonal antiserum and a second antibody labelled with colloidal gold.(x100.000)

On the contrary, the colloidal gold labelled to TF
receptors was not present into such cells, but only
in the primary and secondary spermatocytes and in
early spermatids. Concerning the specimens prepared
with aIR-3 to human Somatomedin C receptor, a positive reaction was observed in spermatocytes I and
II and early spermatids (Fig.4 A,B).

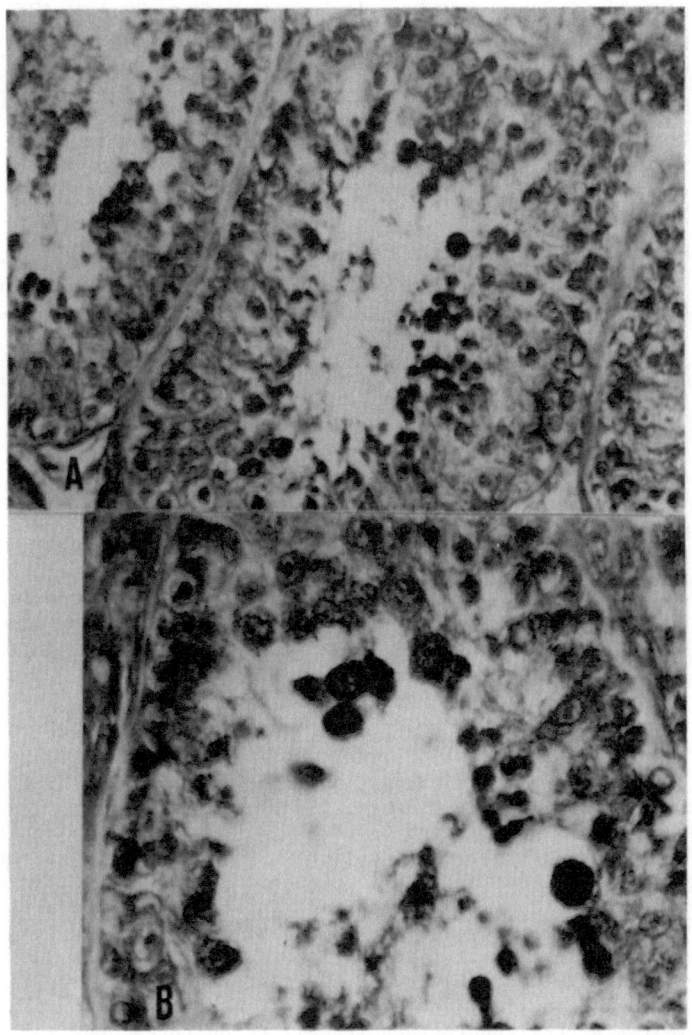

Fig.4 A,B. Sections of a human testicular tubule (Immuno-
          peroxidase Technique), treated with monoclo-
          nal Somatomedin C receptor antibody aIR-3.
          Round cells interpreted as secondary sperma-
          tocytes and early spermatids are stained.

DISCUSSION

According to our results, TF is immunologically detectable in Sertoli cells. A slight positive reaction may be obtained in tubules with normal spermatogenesis at the level of spermatocytes and early spermatids. This positive reaction may be interpreted as due to the receptor TF complexes.
On the other hand TF receptor (positive for the B3/25 glycoprotein) and Somatomedin C receptor (positive for the aIR-3) were found only in spermatocytes and early spermatids. TF receptors were not found in Sertoli cells and spermatozoa present in the tubules.

Conclusions

Our findings in human seminiferous tubules seem to demonstrate that Sertoli cells are devoted to the production of TF and Somatomedin-like peptides, whereas spermatocytes and early spermatids probabily utilize these proteins for their processes of proliferation and differentiation.

ACKNOWLEDGEMENTS

This paper was supported by grant (n. 120103459) from University of Florence. The authors thank Dr. S. Jacobs (The Wellcome Research Triangle Park, NC 27709, U.S.A.) for the very kind gift of Somatomedin C receptor antiserum (aIR-3).

REFERENCES

1. Mather J.P., Gunsalus G.L., Musto N.A., Cheng C.Y., Parvinen M., Wright W., Perez-Infante V., Margioris A., Liotta A. Becker R., Krieger D.T. and Bardin C.W. The hormonal and cellular control of Sertoli cell secretion. J. Steroid. Biochem. 1983: 19, 41
2. Holmes S.D., Lipshultz L.I. and Smith R.G. Regulation of Transferrin secretion by human Sertoli cells cultured in the presence or absence of human peritubular cells. J. Clin. Endocrinol. Metab. 1984: 1058,59
3. Trowbridge I.S. and Lopez F. Monoclonal antibody to transferrin receptor blocks transferrin binding and inhibitis human tumor cell growth "in vitro". Proc. Natl. Acad. Sci. USA 1982: 79,1175
4. Smith E.P., Van Wyk J.J., Svoboda M.E., Kierszenbaum A.L. and Tres L.L. Cell binding specifity and biochemical characterization of somatomedin-C in rat Sertoli-spermatogenic cell co-cultures. Meeting of the American Society of Cell Biology, Kansas City, MO, November, 1984
5. Sylvester S.R., Griswold M.D. Localization of transferrin and transferrin receptors in rat testes. Biol. Reprod. 1984: 31, 195
6. Sternberger L.A. In: immunocytochemistry, New York, Wiley Medical Publication, John Wiley and Sons, 1979 p. 104

Immunocytochemistry of the extracellular matrix (ECM) in normal human testis

R. ZANCHETTA, P.F. MUNARI

Institute of Human Anatomy, University of Padua, Padua, Italy

Introduction

The human seminiferous tubules are composed by an epithelial basement membrane (BM) whose major components are collagen IV and laminin
and by two to six alternating layers of peritubular myoid cells separated by collagen and amorphous ground substances. In the intact testis
Sertoli cells and peritubular cells are adjacent to each other in the
boundary tissue of the seminiferous tubules separated by a basal lamina. Type IV collagen and laminin are two major constituents of BM.
Other components include heparan sulfate proteoglycan and fibronectin.

Some recent data indicate that extracellular matrix (ECM) may play
an important role in seminiferous tubules morphogenesis and germ cells
differentiation: Hadley et al. (1985) suggested that ECM can modulate
the morphogenesis and function of Sertoli cells in culture while Skinner
et al. (1985) demonstrated that peritubular cells and Sertoli cells in
monoculture produce different ECM components, both qualitatively and
quantitatively, and u most of the ECM components appear to remain in
the soluble form in monoculture maintained in serum free medium.

Because a number of human spermatogenic disorders is known to induce
some alterations of ECM we investigated the distribution of the ECM components in normal human testis in an attempt to further investigate the
alterations in some diseases of the testis.

Materials and methods

Frozen unfixed sections (4 μm) of normal human testes, obtained from
two patients with prostate carcinoma, not treated with oestrogens, were
reacted on antibodies solution in PBS: anti-mouse laminin at 1:16; anti-
human fibronectin at 1:8; anti-human type IV collagen at 1:16; anti-
heparan sulfate proteoglycan at 1:10 and anti-tropoelastin at 1:4

using the traditional indirect immunofluorescent technique and a new immunoperoxidase procedure based on the properties of biotin/avidin system.

Results

Both type IV collagen and laminin were localized as an intense band to the basal lamina of seminiferous tubules and of interstitial blood vessels (Fig.1A-B) while they showed a weak staining on myoid cells of tubular wall.

The fibronectin diffusely stained the interstitium, the endothelium of the blood vessels and the myoid cells layers (Fig.1C-D).

The tropoelastin showed a filamentous distribution of staining on the tubular wall and on the blood vessels (Fig.1E-F).

The heparan sulfate proteoglycan diffusely and weakly stained the interstitium and the tubular walls while it showed a strong reactivity on the blood vessels (Fig.1G-H).

Discussion

Most cells in organisms are in contact with an intricate meshwork of interacting, extracellular macromolecules that constitute extracellular matrix. These molecules are secreted locally and assembled into an organized meshwork in the extracellular space of most tissues. Now it is clear that the matrix plays a far more active and complex role in regulating the behaviour of the cells that contact it influencing their development, migration, proliferations, shape and metabolic function.

The BMs are thin layers of specialized ECM that underlie all epithelial cells and they seem to be able to induce cell differentiation, influence cell metabolism and organize the proteins in adjacent plasma membranes: the distribution of laminin and type IV collagen, and probably of heparan sulfate proteglycan, on BMs of seminiferous tubules of normal human testis suggest that in vivo Sertoli cells synthesize both lamin, type IV collagen and heparan sulfate while they are secreted only weakly by peritubular cells. Indeed we found an intense band to the basal lamina of seminiferous tubules with both laminin and type IV collagen while they showed a weak staining on myoid cells.

The fibronectin, non-collagen glycoprotein of the ECM, was localized on interstitium and on a layer between the basal lamina and the first myoid cells layer: it may be in accord with Skinner et al.(1985) that found in monoculture of peritubular cells a strongly positive immunofluorescent reaction to fibronectin antibody and in co-culture with purified Sertoli cells the immunofluorescence was more intense than that visualized in monocultures of peritubular cells. Therefore,

56

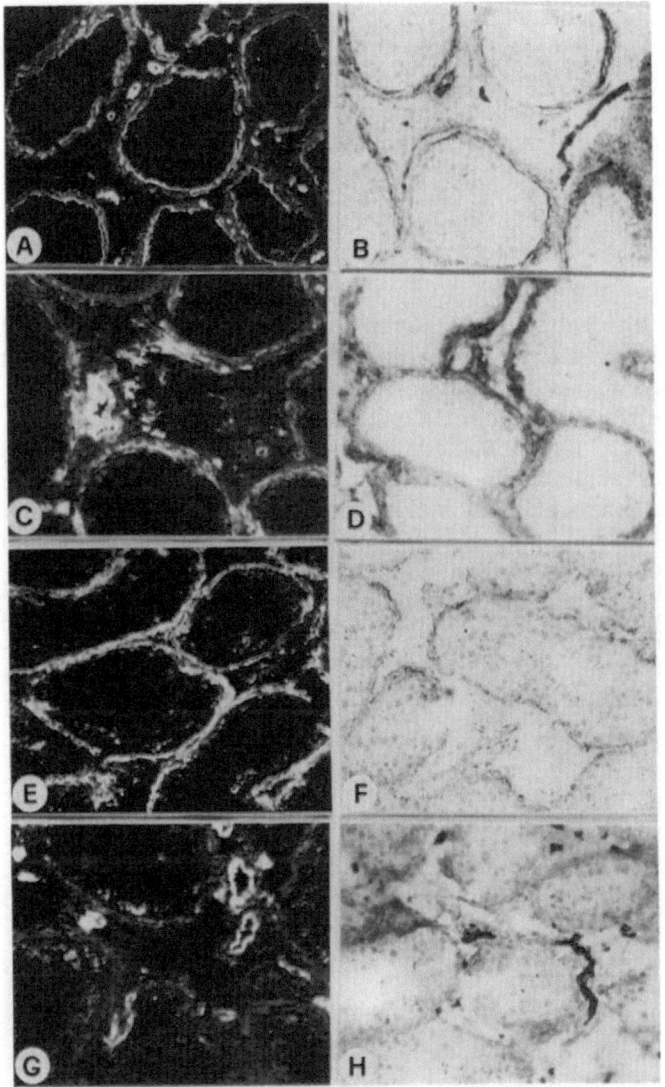

Fig.1. Indirect immunofluorescence and immunoperoxidase technique on frozen unfixed sections
of normal human testis reacted on laminin and
type IV collagen (A-B), fibronectin (C-D), tropoelastin (E-F) and heparan sulfate proteoglycan
(G-H). (x 140)

we can think that also in vivo the secretion of fibronectin may be
done both by peritubular cells and in co-operation with the Sertoli
cells giving the characteristic deposition as we observed in normal
human testis.

In conclusion,these matrix proteins are thought to interact and to assemble in a large variety of different structures, ordered in part by the cells secreting the matrix. Moreover we can hypothesize that, as observed in other tissue (Farquhar M.G.,1982), the ECM components play a role in forming the selective cellular barrier of the testis regulating the passage of macromolecules from the blood to the testis.

References
1. Hadley M.A., Byers S.W., Suarez-Quian C.A., Kleinman H.K., Dym M. Extracellular matrix regulates Sertoli cells differentiation, testi cular cord formation, and germ cell development in vitro. J. Cell Biol. 101, 1511-1522, 1985.
2. Skinner M.K., Tung P.S., Fritz I.B. Cooperativity between Sertoli cells and testicular peritubular cells in the production and deposition of extracellular matrix components. J. Cell Biol. 100, 1941-1947, 1985.
3. Farquhar M.G. The glomerular basement membrane: a selective macromolecules filter. In:Cell Biology of Extracellular Matrix. (E.D. Hay, ed.). pp 335-378. New York, Plenum Press, 1982.

The biological cycle of the interstitial cells in the human testis

G. TEDDE, A. MONTELLA

Institute of Normal Human Anatomy, University of Sassari, Sassari, Italy

Some years ago the hypothesis was suggested that the interstitial cells of the testis undergo a well defined cellular cycle, characterized by three phases of activity: differentiation, maturity and depletion (1, 2, 3).

With the aim of confirming and bearing out the hypothesis, testicular biopsies were analyzed both in nor_ mal and infertile adult individuals.

Material and Methods

Bioptic fragments of testes of 74 patients, aged from 16 to 64, with normal or impaired spermatogenesis (due to idiopathic Sertoli-cell-only syndrome, Klinefelter's syndrome, cryptorchidism, varicocoele, long-term estrogen therapy, oligozoospermia, defects of spermatozoal motility) were utilized.

The fragments were fixed in glutaraldehyde 1.7% and osmium tetroxide 1.0% both in phosphate buffer 0.1 M pH = 7.4, dehydrated in ethanol and propylene oxide and embedded in Durcupan ACM (Fluka AG).

The pieces were cut with the ultramicrotomes LKB "Ultrotome III" and Reichert "Ultracut". Semithin sections 1.0 - 1.5 um thick, after removal of the resin by Na me- thoxide, were stained with toluidine blue for a preliminary study at the light microscope. Ultrathin sections, stained with uranyl acetate and lead citrate, were observed using electron microscopes Siemens Elmiskop 1 A and Zeiss EM 109 (the last in the EM Centre of the University of Sassari).

Results
        The results confirm quite that, on the basis of
the structural and ultrastructural characteristics, three
main patterns of the Leydig cells can be recognized both in
normal and pathological human testes.
        Under the light microscope, the semithin sections
show (a) clear cells   with a finely granulated cytoplasm in which Reinke crystals can be seen; (b) cells
showing a marked basophily and (c) elements characterized
by the presence, in their cytoplasm, of vacuoles more or
less numerous.
        Under the electron microscope, in the interstitial
compartment of the male gonads, it is possible to recognize
numerous cells showing the submicroscopic characteristics
of steroids producing elements. Their cytoplasm presents a
well developed smooth endoplasmic reticulum, mitochondria
with tubular cristae and lipidic inclusions (Figs. 3, 4);
neighbouring cells are joined by wide linear (Fig. 3, arrow)
and anular gap-junctions; in the cytoplasm Reinke crystals
or other crystalline inclusions can frequently be seen
(Fig. 4, arrow). Among the steroid producing cells, others
were observed showing a cytoplasm which is very rich in free
ribosomes and furnished with cisternae of rough endoplasmic
reticulum and mitochondria having linear cristae (Figs. 1,
2). Cells showing submicroscopic characteristics intermedia
te between the above described cells and the steroids producing ones can also be observed (Fig. 2, arrow).
        At last, in the interstitial compartment cells
showing marked phenomena of involution can be seen; they
are characterized by the presence, in their cytoplasm, of
numerous vacuoles, lysosomes and autophagic bodies (Figs.5,
6).

Discussion and Conclusions
        The obtained results confirm quite the prelimina
ry observations obtained in men (1, 3) and experimentally
in rabbits (2). In most of normal and pathological testes
three phases of the Leydig cells cycle can be recognized,
i.e. differentiating cells which develop from undifferentiated fibroblast-like elements; through a "blastic phase"
(4, 5), during which they present abundant rough endoplasmic
reticulum and ribosomes (basophilic cells), develop in
active steroidogenetic cells (clear cells). These last conclude the active endocrine life undergoing a depleting cell
phase characterized by an intensive vacuolisation of the
cytoplasm.

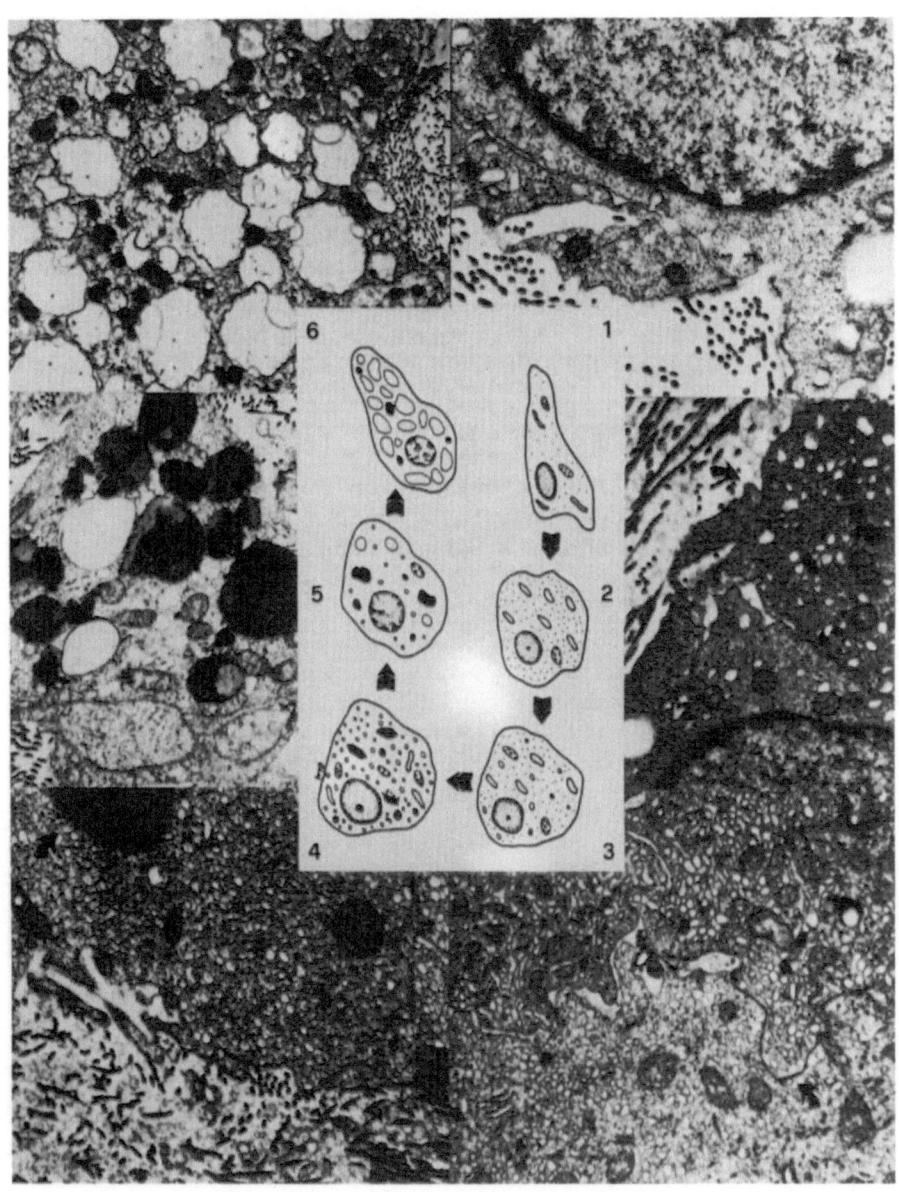

The drawing shows the main steps of the human Leydig cells
cycle. The corresponding submicroscopic patterns can be
seen in the electronograms. 1, 2: differentiating cells
(the arrow indicates an intermediate phase); 3, 4: active
steroidogenetic cells (arrows indicate a gap-junction and
a Reinke crystal); 5, 6: depleting cells.

61

Likely, many factors play a role in the control of the biological cycle of the human Leydig cells, i.e. gonadotropic hormones, active substances able to modulate the gonadal interstitial microenvironment and the influence of the vegetative nervous system (6).

References
1) TEDDE G, ZECCHI S. Caratteristiche microscopiche e sub-microscopiche delle cellule interstiziali del testicolo umano in individui azoospermici ed oligospermici. Arch. It. Anat. Embriol. 1977: suppl. 82, 357-58.
2) TEDDE G, MORABITO A, MONTELLA A, TEDDE R, PALA AM, TEDDE PIRAS A. Risultati preliminari di uno studio sperimentale sul ciclo biologico delle cellule interstiziali del testicolo. Studi Sassaresi 1981: 59, 101-31.
3) TEDDE G, MONTELLA A. L'évaluation du cycle des cellules de Leydig humaines dans la biopsie testiculaire. C.R. 1er Congr. Soc. Androl. Lang. Franç., Lyon, 1983.
4) BALBONI GC. The problem of ovarian stroma. Arch. It. Anat. Embriol. 1973: 78, 37-58.
5) BALBONI GC. Histology of the Ovary. In: The endocrine function of human ovary, eds. James VHT, Serio M, Giusti G. New York, Academic Press, 1976:1-24.
6) TEDDE G, TEDDE PIRAS A. L'innervation du testicule humain étudiée par la microscopique électronique. Bull. Ass. Anat. 1984: 200, 77-86.

Does bovine follicular fluid stimulate Leydig cell steroidogenesis?

G.P. RISBRIDGER, S. AVERILL, D.M. DE KRETSER

Department of Anatomy, Monash University, Clayton, Victoria, Australia

The role of pituitary hormones in the regulation of gonadal function is well documented. However there are substantial data to demonstrate that local factors, in addition to pituitary gonadotrophins, can regulate gonadal function. Studies of the paracrine regulation of testicular function have documented numerous factors emanating from the tubules, macrophages or interstitial fluid which effect Leydig cells to alter steroidogenesis in vitro (see review by Findlay & Risbridger, 1) (1). In addition to Sharpe and Cooper (2) our own Laboratory has reported the presence of activity in testicular interstitial fluid which stimulates testosterone production by Percol purified Leydig cells in vitro. The addition of interstitial fluid causes a linear dose dependent rise in testosterone production in the absence or in the presence of a maximally stimulating dose of hCG (3). Based on the observations that the ovary and testis often produce the same peptides, (for example, inhibin) the aim of the study was to determine whether or not bovine follicular fluid contains activity which stimulates Leydig cells to steroidogenesis in vitro.

Insterstitial fluid was collected as previously described (2) from adult male rats. Briefly, an incision was made in the tunica albuginea nd the testis was suspended and the fluid was allowed to drain and collect overnight at 4°C. The interstitial fluid was then charcoal treated to remove an endogenous steroids prior to assay. Bovine follicular fluid, free of cystic fluid, was aspirated within 30 minutes of collection of ovaries from the local abbattoir. The bFF was then charcoal treated as described for the interstitial fluid. The Leydig cells were obtained by collagenase digestion of adult male rat testis. Purified Leydig cells were harvested after centrifugation of interstitial cells through Percoll density gradients. The purified Leydig cells were then incubated with test substances, in triplicate, at multiple dose levels, for 20 hours at 32°C. Testosterone or pregnenolone production was then measured by specific radioimmunoassay.

Interstitial fluid and bovine follicular fluid caused linear dose dependent stimulation of testosterone production by Percoll purifief Leydig cells in vitro. In the presence of maximally stimulating doses of hCG and hydroxy cholestorol, the addition of interstitial fluid or bovine follicular fluid did not cause any further increase in testosterone production by the Leydig cells. However if pregnenolone production was measured in the same samples there was a linear dose dependent rise in pregnenolone production. These results would indicate that the interstitial fluid and the bovine follicular fluid were able to stimulate steroidogenesis at a step prior to the production of pregnenolene.

Microscopic examination of the cells cultured either in the absence or in the presence of hCG showed little difference in histological appearance. The familiar features of Leydig cells are evident; the nucleus contains the peripheral rim of heterochromatin and the cytoplasm contains abundant mitochondria, typical of these steroid secreting Leydig cells. Despite the fact that testosterone concentration is increased almost ten fold, there is little difference in the appearance of the Leydig cells cultured with or without hCG. However in the presence of interstitial fluid the microscopic appearance of the Leydig cells is dramatically altered. These cells have flattened on to the surface of the dish and their appearance is similar to that observed after culture in bovine follicular fluid.

These data therefore demonstrate that bovine follicular fluid stimulates steroidogenesis in vitro by Percoll purified Leydig cells. The stimulatory activity appears to be similar to that present in the testicular interstitial fluid which suggests that the factor is present in these two biological fluids. However definitive proof of this fact requires the identification or the purification of the factor responsible for that activity in the respective fluids.

REFERENCES

1   Findlay, J.K. & Risbridger, G.P. Intragonadol Control Mechanisms. in: Clinics in Endocrinology & Metabolism – Reproductive Endocrinology. Eds. Burger, H.G., Saunders, W.B., U.K. 1987 (In Press).

2. Sharpe, R.M. & Cooper, I. Intratesticular secretion of a factor(s) with major stimulatory effects on Leydig cell testosterone secretion *in vitro* Mol. Cell Endocrinol. 1984 : 37, 159-168.

3. Risbridger, G.P., Jenkin, G. & de Kretser, D.M. The intraction of hCG, hydroxysteroids & interstitial fluid on rat Leydig cell steroidogenesis. J. Reprod. Fert. (1986). 77 : 239-245.

Testicular ultrastructural modifications during pharmacological treatment in patients affected by prostatic cancer

M. BOLOTTI, L. MEDOLAGO-ALBANI[1], F. DI SILVE-RIO[2], R. TENAGLIA[2], A. FIORENTINI, G. SPERA

Clinica Medica V; [1]Dipartimento Biopatologia Umana; [2]Patologia Urologica, Università La Sapienza, Roma, Italy

INTRODUCTION

At present, pharmacological therapy of androgen-dependent prostatic cancer is based upon the use of antiandrogen drugs (i.e. Cyproterone acetate or Flutamide) (1,2) or LHRH analogues (3,4), alone or combined (5). All of these substances act also at testicular level inducing morphofunctional alteration of spermatogenesis and steroidogenesis.
In this view, the present study was designed to evaluate the effect of Cyproterone acetate (CPA), Flutamide (FL), Zoladex (ZO) and CPA+ZO on interstitial tissue and germinal epithelium in patients with hormonal dependent prostatic cancer.

MATERIALS and METHODS

Thirty patients with a diagnosis of androgen-dependent prostatic cancer have been submitted to treatment with CPA (200mg/die per os) or FL (750mg/die per os) or ZO (3.6mg every 28 days subcut) or CPA+ZO (at same doses) for a variable period ranging from three to six months. Before and after the treatment the same subjects underwent testicular biopsy and the fragments obtained were examined at light and electron microscope to better investigate the morphological status. Radioimmunoassays of FSH, LH and testosterone were performed for hormonal study.

RESULTS

As LH, FSH and testosterone levels are concerned, our data are in agreement with the results of other authors. In particular, using FL testosterone blood values were between the normal range; on the contrary gonadotrophin levels appeared to be increased despite the results obtained in the ZO and CPA plus ZO groups.
Antiandrogen drugs CPA and FL have different effects on testicular tissue. In fact, the former induces considerable damage in the inter-

stitium (fig.1) which becomes poor in Leydig cells morphologically
normal. Collagene is as clearly represented as mesenchymal cells and
mast-cells. In apparent contrast with interstitium, the germinal epithelium matured regularly up to the late spermatids stage even if the
Sertoli cells presented generic signs of degeneration (fig.2). As far
as the FL is concerned, its effect on Leydig cells was markedly different from that of the CPA. In particular Leydig cells were apparently normal and typical showing some irregularity only in the cellular
envelope that appeared digitated (fig.3). The smooth endoplasmic reticulum and mitocondria seemed overdeveloped as in case of hyperstimulation. On the other hand, the germinal epithelium showed a complete
spermatogenesis, but on the late spermatids it can be noted golgi complex hypertrophy and severe acrosomal anomalies.
The effect of ZO was also studied. The germinal epithelium showed a
progressive arrest. The most significant alterations were in the A-type
spermatogonia, in fact foldings of the two layers of nuclear envelope
including chromatin or cytoplasm were present (fig.4). This particular
picture was found only in the specimens obtained after the treatment.
In the cytoplasm, fragments of nuclear membrane surrounding cytoplasm
together with well-established nuclear pores were noted (fig.5).
The interstitial tissue presented a marked de rrangement. It was characterized by the presence of abundant connective tissue and mesenchymal cells and the rare Leydig cells were greatly altered showing lipidic and lipofuscinic degeneration (fig.6).

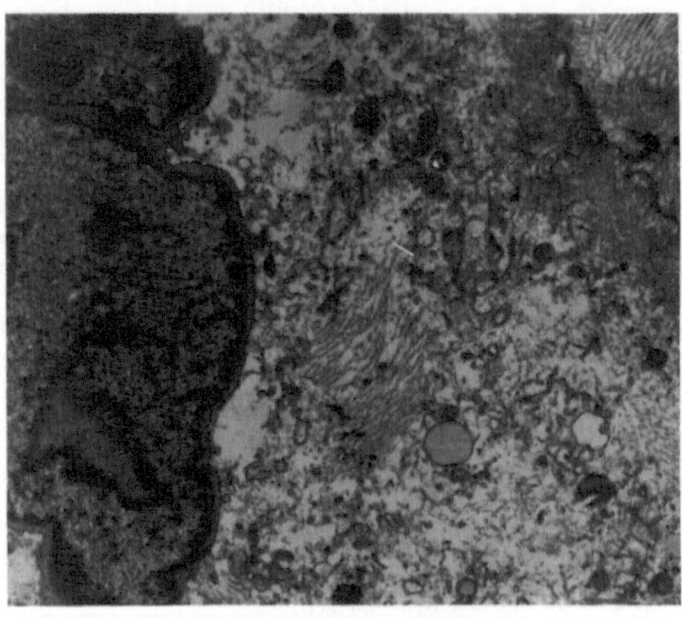

Fig.1. Electron photomicrograph of Leydig cells.

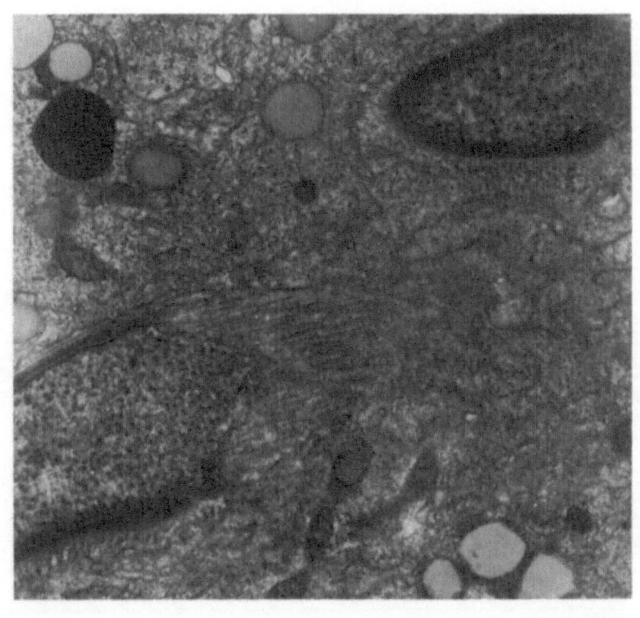

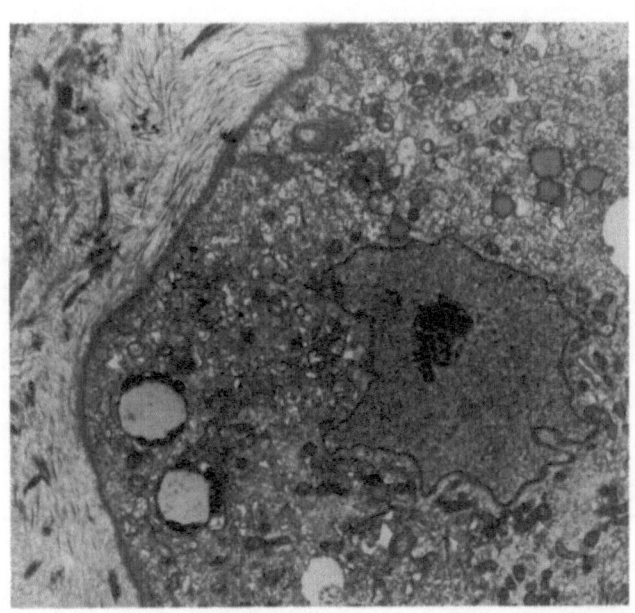

Fig.2. Electron photomicrograph of Sd-type spermatids and Sertoli
cell.

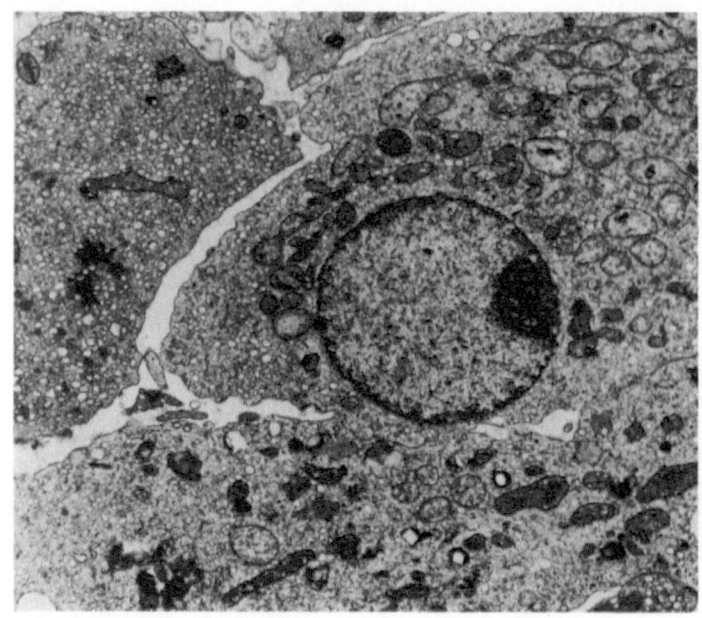

Fig.3. Electron photomicrograph of Leydig cell: smooth reticulum and
mitocondria appear digiteated.

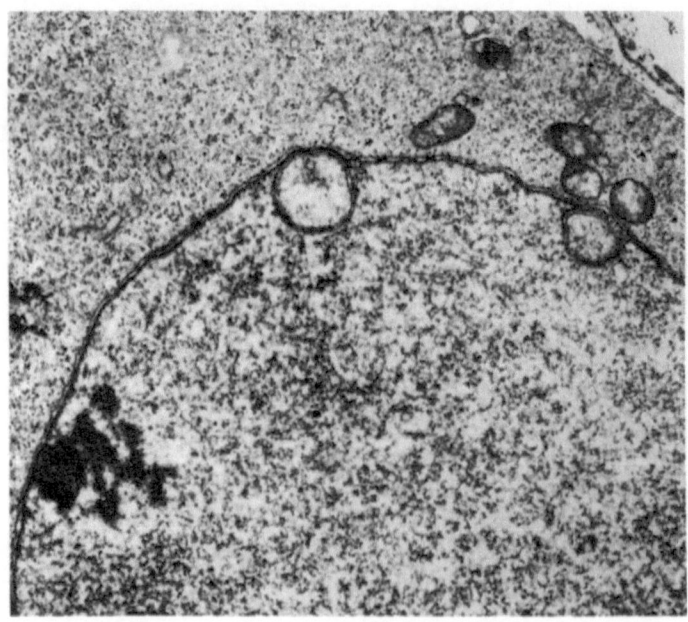

Fig.4. Electron photomicrograph of A-type spermatogonia showing the
foldings of the two layers of nuclear envelope.

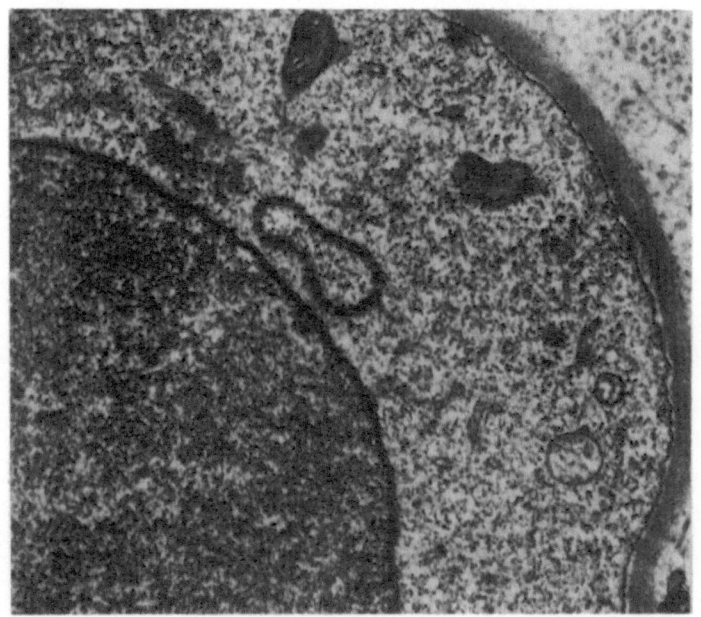

Fig.5. Electron photomicrograph of spermatogonia showing a fragment of
nuclear envelope surrounding cytoplasm. Nuclear pores are evident.

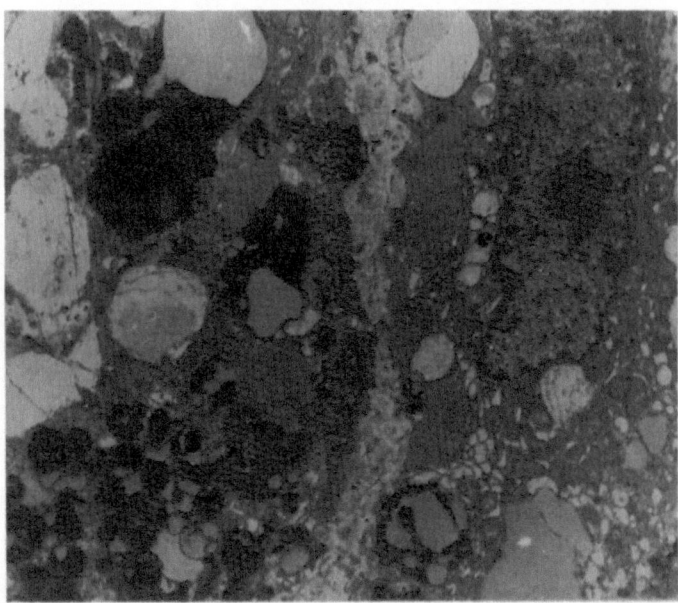

Fig.6. Electron photomicrograph of Leydig cell showing lipidic and lipofuscinic degeneration.

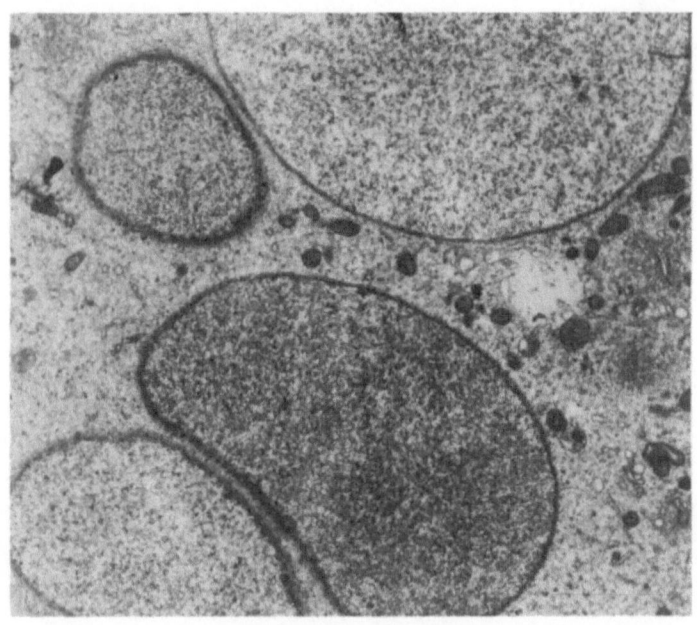

Fig.7. Electron photomicrograph of spermatogonial plurinucleated cell
showing evident nuclear pores.

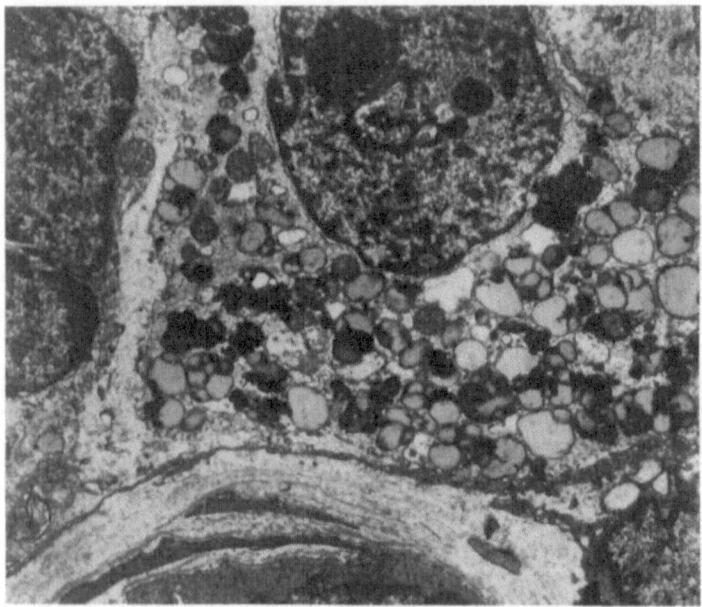

Fig.8. Electron photomicrograph of interstitial tissue showing degenerated Leydig cells.

70

With the association of CPA plus ZO we observed either the presence of spermatogonial plurinucleated cells (fig.7) and a spermatogenesis arrest, together with lipidic and lipofuscinic degeneration of Leydig cells that appeared to be more represented than in the CPA treatment alone (8).

DISCUSSION and CONCLUSIONS

The data obtained in the present study lead us to consider CPA as acting directly on Leydig cells blocking irreversibly both spermatogenesis and steroidogenesis (6).
On the contrary, FL, acting at a receptor site, did not interfere directly with the testicular functional morphology (7).
The effect of the LHRH analogue (ZO) is on steroidogenesis, inducing peculiar damages on Leydig cells and reducing the Leydig cells / undifferentiated mesenchymal cells ratio. The effect on germinal epithelium seemed to be highly specific; in fact the ZO caused spermatogesis arrest with a selective action on A-type spermatogonia.
It can be pointed out that changes in the nuclear envelope and nuclear pore complex are characteristic of this treatment and are similar to those observed in spermatogenesis arrest due to a selective reduction of cellular clones as in some idiopathic maturative arrest (8) and in cryptorchidism (9), thus leading to hypothesize a drug induced effect, or a spontaneous one.
The association of CPA plus ZO interestingly seemed to protect testicular interstitium. In fact the degree of damages being found in Leydig cells is less pronounced with respect to both treatment alone.
In conclusion, all these substances have been proposed as possible antifertility agents, depending on their actions (direct or indirect) on spermatogenesis, although the reversibility of the drug effect at the end of the treatment has to be proved.

REFERENCES

1) Neuman F. Pharmacology and potential use of Cyproterone acetate. Horm Metab Res 1977: 9; 1-13.

2) Aihrart RA, Barrult TF, Sullivan JW, Levine RL, Schlegel JU. Flutamide therapy for carcinoma of the prostate. South Med J 1978: 71; 798-91.

3) Santen RJ, Warner B, Demers LM, Dufau M, Smith J. Use of GnRH hormone agonists analogues. In : LHRH and its analogues - a new class of contraceptive and therapeutic agents. eds. Vickery B, Nestor JJ Jr, Hafez ESE. Boston MTP Press, 1984.

4) Ahmed SR, Shalet SM, Brooman PJC, Howell A, Blacklock NJ, Rickards D. Treatment of advanced prostatic cancer with LHRH analogue ICI 118630: clinical response and hormonal mechanism. The Lancet 1983.

5) Labrie F, Bélanger A Dupont A, Lefebvre FA, Labrie C, Lacourciére Y, Raynand JP, Husson JM, Emond J, Houle JG, Girard JG Monfette G, Pasquet JP, Valliéres A, Bossé C, Delisle R. Combined anti-hormonal treatment in prostate cancer, a new approach using an LHRH agonist or castration and an antiandrogen. In: Hormones and Cancer. New York, Raven Press, 1984.

6) Spera G, Di Silverio F, Medolago-Albani L, Coia L, Di Cicco G, Tenaglia R, Stigliani V. Microscopia elettronica delle variazioni indotte sui testicoli da un pluriennale trattamento con ciproterone acetato. In: Andrologia, eds. Gelmini. Milano, 1983: 133-39.

7) Labrie F, Dupont A, Belanger A, St-Arnaud R, Giguère M, Lacourciere Y, Emond J and Monfette G. Treatment of prostate cancer with gonadotropin-releasing hormone agonist. Endocrine Reviews 1986: 1; 67-74.

8) Fabbrini A, Francavilla S, Bellocci M, Moscardelli S, Martini M, Bruno B. Gli arresti maturativi della spermatogenesi. In: Andrologia, eds. Piccin. Pisa, 1981: 13-56.

9) Osculati F, Amati S, Petrini E, Caucci M. Considerazioni sulle lesioni iniziali del testicolo umano nel criptorchidismo. Rassegna Italiana di Chirurgia Pediatrica 1982: 2; 92-96.

Ultrastructural aspects of retractile testis: a preliminary study

S. CINTI, E. PETRINI, S. AMATI, F. DI GIACINTO[1], D. LOMIENTO[1], M. CAUCCI[1]

Istituto di Morfologia Umana Normale, Università di Ancona; [1]Chirurgia Pediatrica, Ospedale G. Salesi, Ancona, Italy

INTRODUCTION

Numerous morphological and clinical studies have been carried out on cryptorchid testes by various authors (1,2, 3,4,5,6) and a series of anatomo-clinical classifications (7,8) has been proposed with a view to providing precise indications for adequate therapies (9).
For some years our group has been working on normal and cryptorchid testes, concentrating on ultrastructural aspects (10,11,12). The frequent occurrence of abnormal retractile testes and ectopic inguinal testes resulting from retractile testis directed our research towards a morphological examination of these organs.
This study presents our preliminary ultrastructural data. We analyzed the retractile testis biopsies of two groups of patients: 5 subjects underwent surgery at age 7, (first group), 7 at age 11 (second group); results were compared to data obtained from normal and cryptorchid testes of patients of the same age.

MATERIALS AND METHODS

After surgery biopsies were fixed in 2% cacodilate buffered solution of glutaraldehyde, then postfixed in 1% osmium tetroxide and embedded in a mixture of Epon-Araldite. Ultrathin sections were examined with Philips Electron Microscope 301 or Zeiss 902 Electron Microscope.

RESULTS

Ultrastructural observation revealed same analogies between the altered morphology of seminiferous tubules in retractile testis and that observed in cryptorchid patients, even accounting for the wide range of normal individual variations.

73

Seven-year-old patients (first group)

Spermatogonia
These retractile testis biopsies presented some binucleate spermatogonia.

Sertoli cells
They presented evident aspects of immaturity: roundish nuclei, few cytoplasmic organelles.

Basal membrane
The basal membrane of seminiferous tubules in retractile testis showed a slight focal thickening (Fig.1).

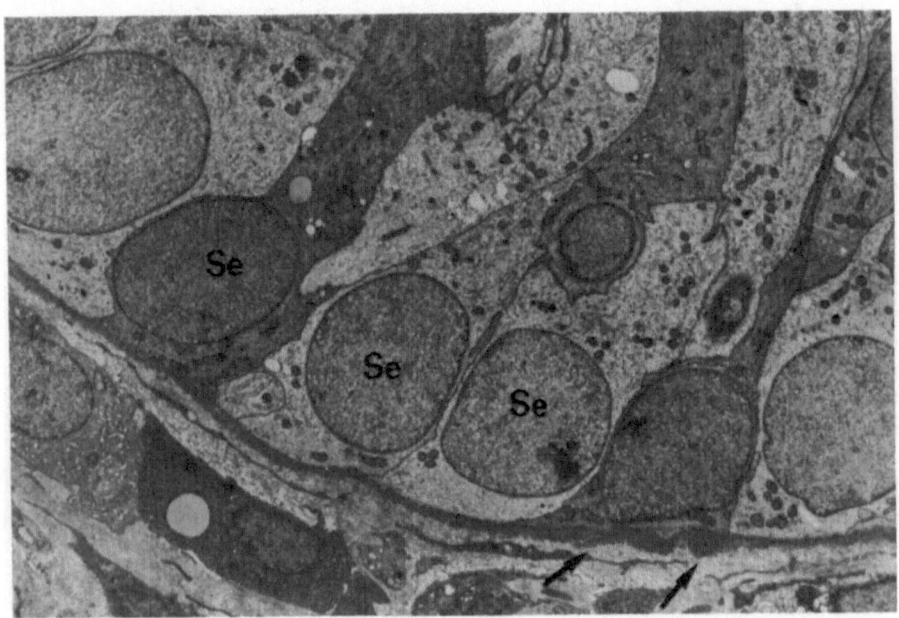

Fig.1.   Section of a seminiferous tubule of retractile testis in a 7 years patient. We can observe normal Sertoli cells (Se). The basal membrane is slowly thickened (➤) (17.750 X).

Eleven-year-old patients (Second group)

Spermatogonia
Whereas normal testes of this age group present all the steps of germ cell maturation (from spermatogonia to spermatids), most retractile testis in our study presented only spermatogonia A pale. Furthermore, their nuclei often showed large blebs of nuclear membrane.

74

<u>Sertoli cells</u>
At age eleven these cells were still markedly immature,
with roundish nuclei and scarce R.E.R. Often there were
large lipofuscinic deposits in the cytoplasm and at times
they were especially abundant (Fig.2).

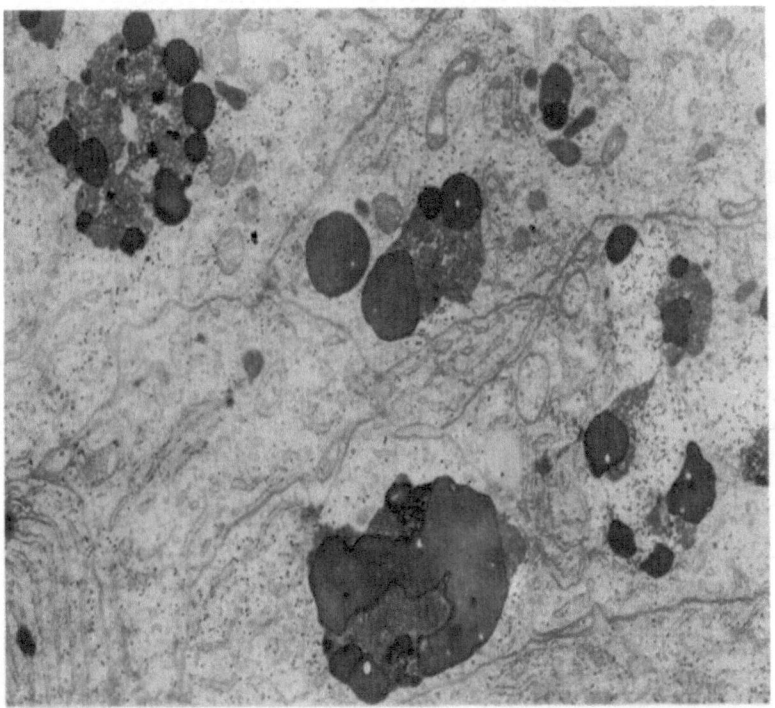

Fig.2. Sections of cytoplasm of Sertoli cells of a seminiferous tubule in an 11 years patient. We can
observe the presence of large lipofuscinic deposits. (17.750 X).

<u>Basal membrane</u>
It was thickened and presented splittings and large knobs
protruding into the cytoplasm of tubular elements.
The inner acellular layer was often thickened and disorga-
nized.

<u>DISCUSSION</u>

Ultrastructural observation of retractile testis biopsies
at age seven showed an initial delay in Sertoli cell evolution, which was more evident and frequent in the eleven-
year-old subjects. The presence, in some patients of the
latter group, of abundant lipofuscinic deposits in the cyto
plasm of Sertoli cells revealed a definite alteration in

75

the metabolism of these cells; similar deposits had already been reported in cryptorchid testis (13). In the seven-year-old group the development of germ cells did not appear to be undermined, however binucleate elements were in excess of normal values, and cryptorchid testis were observed to present even tri- and tetranucleate spermatogonia. Delay in germ cell maturation was more apparent in the second group of patients, where only spermatogonia A pale were present in reduced numbers. The same situation has already been described in cryptorchid testis (14). Retractile testis basal membrane was slightly thickened focally at age 7, at age eleven it showed splittings and knobs comparable to those observed in cryptorchid testis at the same age, which demonstrates that also this morphological structure is affected by the abnormal situation of testes (15,16).

CONCLUSIONS

Results of ultrastructural observation of retractile testes compared with those of cryptorchid testes biopsies of seven and eleven-year-old patients suggest a remarkable similarity between these two pathologic conditions.

REFERENCES

1) Baccetti B, Bigliardi E, Vegni-Talluri M, Soldani P, Renieri T, Selmi MG, De Martino A, Bracci R, Vanni MG. The fine structure of the testis in the cryptorchid man. In: Cryptorchidism, eds. Bierich JR and Giarola A, Academic Press, 1979:91-123.

2) Hedinger Chr. Histopathology of the cryptorchid testis. In: Cryptorchidism, eds. Bierich JR and Giarola A, Academic Press, 1979:29-38.

3) Fabbrini A, Santiemma V, Francavilla F, Moscardella S, Catignani P. L'indagine istopatologica del testicolo nella clinica andrologica. In: Andrologia e Fisiopatologia della riproduzione, eds. Field Educational Italia-Acta Medica.Roma, 1984:175-194.

4) Hadziselimovic F, Girard J, Herzog B. Lack of germ cells and endocrinology in cryptorchid boys from one to six years of life. In: Cryptorchidism, eds. Bierich JR and Giarola A, Academic Press, 1979:129-134.

5) Osculati F, Amati S, Petrini E, Caucci M. Considerazioni sulle lezioni iniziali del testicolo umano nel criptorchidismo. Rass It Chir Ped 1982:XXIV(2); 92-96.

6) Schulze C. Survival of human spermatogonial stem cells

in various clinical conditions. Fortschr Androl 1981:7, 58-68.

7) Nistal M, Paniagua R, Diez-Pardo JA. Histologic classification of undescended testes. Human Pathol 1980:11; 666-674.

8) Bussolati G, Papotti M. Aspetti istopatologici fondamentali della infertilità maschile. In: Andrologia e Fisiopatologia della riproduzione, eds. Field Educational Italia-Acta Medica, 1984: 151-160.

9) Hadziselimovic F, Herzog B, Seguchi H. Surgical correction of cryptorchidism at 2 years: Electron Microscopic and morphometric investigations. J'Pediatr Surg 1975:10 (1); 19-26.

10) Osculati F, Amati S, Petrini E, Franceschini F, Caucci M. Aspects of maturation of seminiferous tubules in the human testis. In: Cryptorchidism, eds. Bierich JR and Giarola A, Academic Press, 1979:141-148.

11) Amati S, Petrini E, Gazzanelli G, Osculati F, Caucci M. Ultrastructural aspects of Sertoli cells in infancy. Boll Soc It Biol Sper 1979:LV; 378-382.

12) Osculati F, Amati S, Petrini E, Marelli M, Caucci M. Nota preliminar de Morfologia microscopica del testiculo retenido. Revista Espanola de Pediatria 1979:XXXV (207-208); 265-270.

13) Schulze C. Sertoli cells and Leydig cells in man. In: Advances in Anatomy and Cell Biology, eds.Springer Verlag.Berlin Heidelberg New York Tokyo, 1984:88; 65-66.

14) Baccetti B, Bigliardi E, Vegni-Talluri M, Soldani P, Renieri T, Selmi MG, De Martino A, Bracci R, Vanni MG. The fine structure of the testis in the cryptorchid man. In: Cryptorchidism, eds. Bierich JR and Giarola A, Academic Press, 1979:91-123.

15) Fabbrini A, Francavilla S, Santiemma V, De Martino C, Francavilla F. Ultrastructural changes in the seminiferous tubule wall and intertubular blood vessels in human cryptorchidism. In: Cryptorchidism, eds. Bierich JR and Giarola A, Academic Press, 1979:69-83.

16) De Kretser DM, Kerr JB, Paulsen CA. The peritubular tissue in the normal and pathological human testis. An ultrastructural study. Biol Reprod 1975:317-324

Advances in the histophysiology of the ovary

G.C. BALBONI

Institute of Human Anatomy, University of Florence, Policlinico di Careggi, Firenze, Italy

Introduction

The structure of the ovary is remarkably variable accor
ding to the age and the functional activity.

From the birth to the post-menopause and particularly
during the fertile life the structural organisation of the
organ changes under the influence of the hypophyseal gonadotropins (FSH, LH, prolactin) and other factors such as
prostaglandins, catecholamines, active substances of endo-
gen and/or exogen source (inhibin, relaxin, transferrin,
etc.) and, in addition, nervous influences and variations
in the blood supply.
The cyclic changes of the ovarian structures due to the go
nadotropins are superimposed to a morphological background
that continuously modifies itself with the progressing age
because of the persistence of not functioning structures
such as advanced atretic follicles, regressing corpora lutea, corpora fibrosa et albicantia and the occurrence of va
scular and stromal changes.

The new methods of investigations such as morphometry,
transmission and scanning electron microscopy, histochemistry, immunohistochemistry together with the classic morphological techniques allowed to obtain a lot of information about the ovarian histophysiology.

Aim of this paper is to present a concise dynamic picture of the ovarian morphology with special emphasis on
the data obtained by our recent research.

General remarks

Two zones can be distinguished in the ovary (Fig.1):
1) a peripheral one, the cortical zone, underlying the
so-called germinal epithelium which covers the organ. In
this zone the ovarian follicles and corpora lutea are contained in the cortical stroma. This latter is essentially
cellular, but all its components (cells, intercellular sub
stance and capillaries) undergo some changes in the different phases of the menstrual cycle. The stroma cells can
give rise to endocrine structures such as the follicular

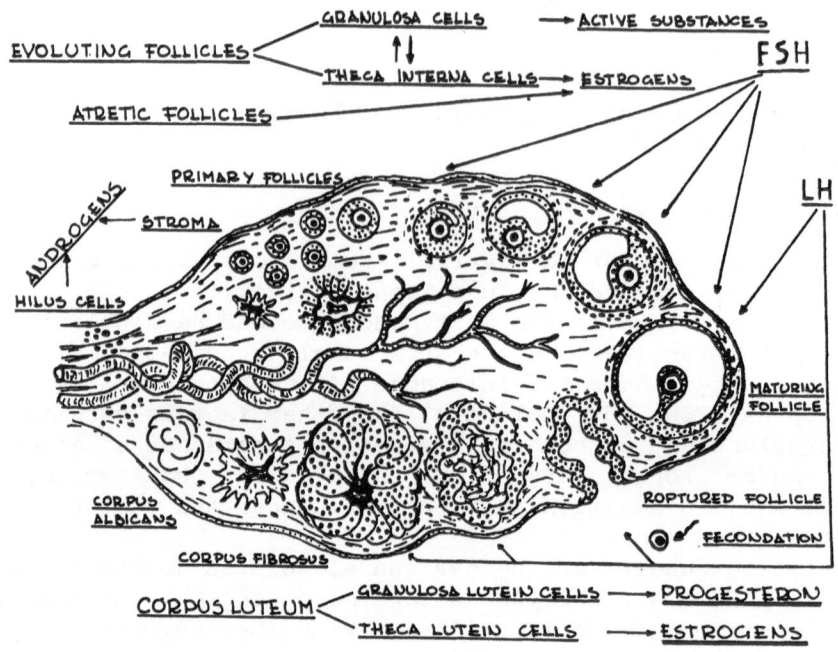

Fig. 1. General view of the ovary's structural organization and of its endocrine activities.
In the medullary zone vascular branches run a spi̱
ral course in loose connective tissue together
with nervous fibers.
In the hilus androgen-producing cells are present.
The cortical stroma contains evolving and regressing follicles and corpora lutea.
At the surface, the so-called germinal epithelium
(after Balboni).

theca interna and the interstitial tissue.
2) the medullary zone is the core of the organ, consisting of loose connective tissue, where a number of blood
vessels run a spiral course. A well developed perivascular
network is present. Near the hilus small groups of cells
(hilus cells) are arranged close to the nervous fibres pe-
netrating from the mesovarium. These cells have structural
and ultrastructural characteristics quite similar to those
of the interstitial cells of the testicle and are supposed
to produce androgens.

80

Three endocrine compartments can be identified in the ovary (Fig.2):

1) the follicle where, in addition to the gametogenic function, the estrogens are produced;

2) the cortical stroma, that produce androgens;

3) the corpus luteum, whose function is to secrete progesteron and some amounts of estrogens.

Other active substances or local hormones are produced, stored up and utilized in the ovary.

The ovarian follicle (Figs 3-4)

In the follicles the gametogenesis and the estrogen production are accomplished.

Follicles may be distinguished as follows:

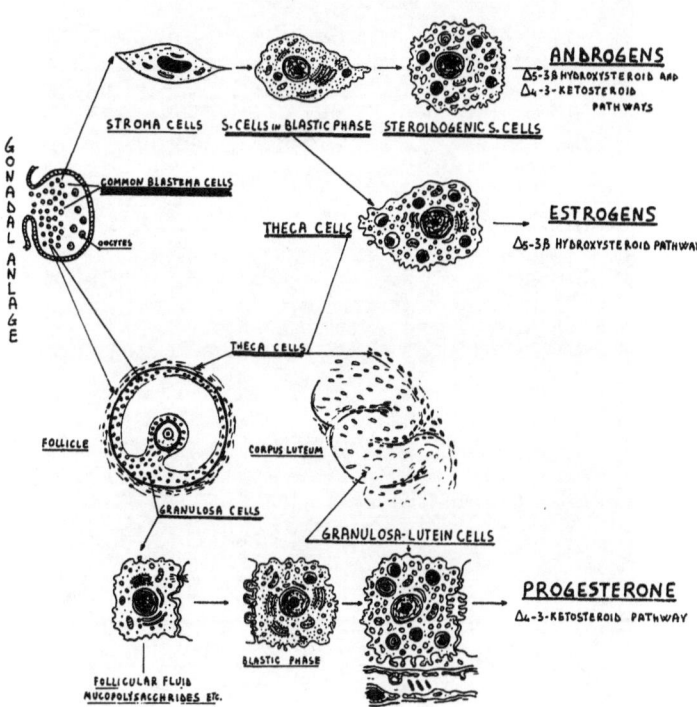

Fig. 2. Diagram showing the different types of endocrine cells in the ovary, their origin from the common blastema, their development and their steroidogenic activity (after Balboni, 1977).

1) <u>Evolutive follicles</u> in different stages of maturation, starting from the primordial follicles. Primary, secondary, cavitary (Graafian follicles) and mature or preovulatory follicles are the main stages considered.
Not all the follicles in a given ovary respond to the gonadotropic hormones and run the entire course of maturation. In the human female, during the fertile life, only around 400 selected follicles ovulate, while the majority degenerate. It was estimated that a human follicle takes about three consecutive menstrual cycles to reach the preovulatory stage.
The selection of the follicle fated to maturation and rupture (dominant follicle) is a very complex phenomenon, where several factors are involved. Inhibin is supposed to play a

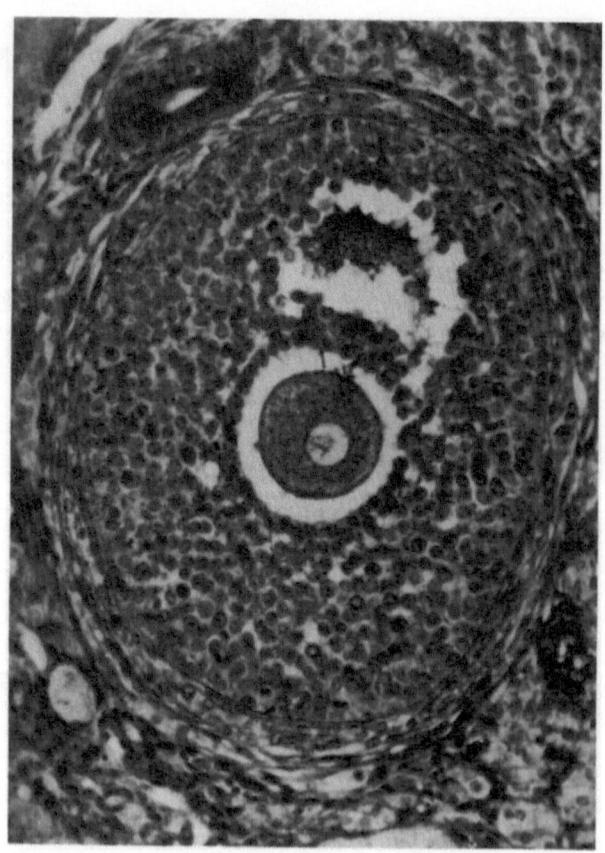

Fig. 3. Initial cavitary follicle. The granulosa cells are actively proliferating (see the numerous mitoses). A theca interna layer begins to deve lop. At the center the oocyte.

82

rôle, blocking the estrogen induced surges of both FSH and
LH. A prominent control is exerted by intraovarian hormones
through paracrine mechanisms. It was observed that estrogen
to androgen ratio in the follicular fluid is high in healthy dominant follicles, while a . high androgen to estrogen
ratio is distinctive of atretic follicles.

   2) Involutive or atretic follicles. The regressive process (atresia) may affect the follicles in all the stages

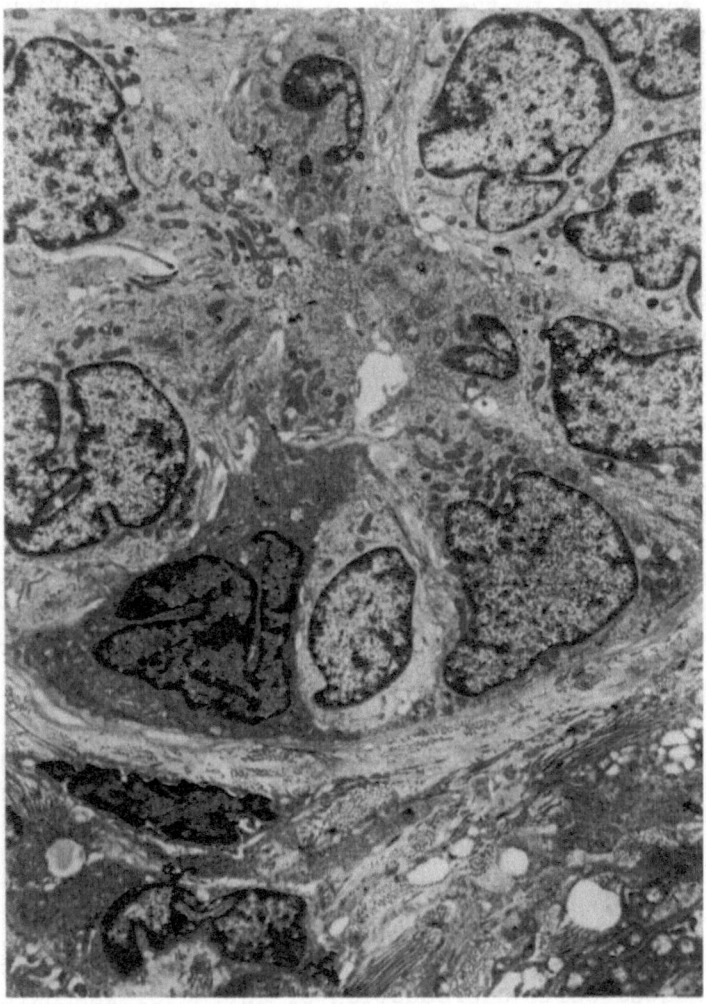

Fig. 4. Electron micrograph of a secondary follicle in
        the human ovary. The follicular cells are rich
        in organelles and their cellular membranes in-
        terdigitate. Some morphological differences
        between these cells are detectable. x 4.800

of their development.

From a morphological point of view, atresia consists essentially of:

1) degeneration of the germ cell, 2) degeneration of the follicular epithelium and its replacement by connective tissue, 3) thickening of the basal membrane that encircles the follicle (atresia membrane), 4) in Graafian involving follicles a hypertrophy of the theca interna occurs.

The atretic process may show different morphological aspects according to the stage of follicular evolution in which it begins. Apart from the primordial and primary follicles, that degenerate without leaving traces, two main types of atresia affect the Graafian follicles 1) the obliterans atresia, affecting the Graafian follicles of small and medium size and leading to the formation of a corpus fibrosus and 2) the cystic atresia that occurs when the follicular cavity is too large for be ing filled up by the connective tissue proliferation.

The granulosa cells of the evolutive follicles (Figs 5-6)

While the rôle of the theca interna cells in producing steroids and particularly estrogens is well ascertained,the problem exists about the functional significance of the granulosa cells. Apart from their activity in favouring the exchanges with the oocyte, during the follicular maturation these cells do not exhibit any morphological appearance of endocrine activity (i.e. steroid production). They have no direct blood supply and all their metabolic exchanges occur by diffusion from/to the theca interna vessels. Their cytoplasmatic equipment consists essentially of a well developed Golgi apparatus, numerous free ribosomes,rough endoplasmic reticulum and small mitochondria with linear cristae. Therefore, they appear to be devoted to protein rather than steroid synthesis. Their protein-synthetic activity is related to several functions such as cellular multiplication and construction of new cytoplasma in granulosa layer, production of the proteins and mucopolysaccharides of the follicular fluid (Balboni, 1976).

In experimental conditions in vitro granulosa cells have been demonstrated to be able to produce a variety of steroids with progesterone as major product, but they failed to produce estradiol. Only in cultures of recombined theca and granulosa cells a production of estradiol was detectable.

In our experience (Balboni and Zecchi, 1981),granulosa cells harvested from large Graafian follicles and cultured in vitro undergo a luteinization process much more evident

84

in presence of HCG in the medium. In vivo, these cells
show the morphological signs of luteinisation and express
their steroidogenic potentiality only when the endocrine,
metabolic, nervous and vascular conditions typical of the
ovulatory phase are realized. Among the proteic products
that granulosa cells can elaborate, an "oocyte meiosis inhi
bitor" was demonstrated. This substance, that is a polypep
tide of low molecular weight, blocks the oocyte in the pro
phase of the first meiotic division, until the LH surge,
and probably prevents the luteinization of the granulosa
cells. LH favours the oocyte maturation and suppresses the
production and/or the activity of the OMI (Fig.7).

Relaxin in the evolutive and luteinizing follicles (Figs 8-9)

Relaxin is a polypeptide hormone, whose production was

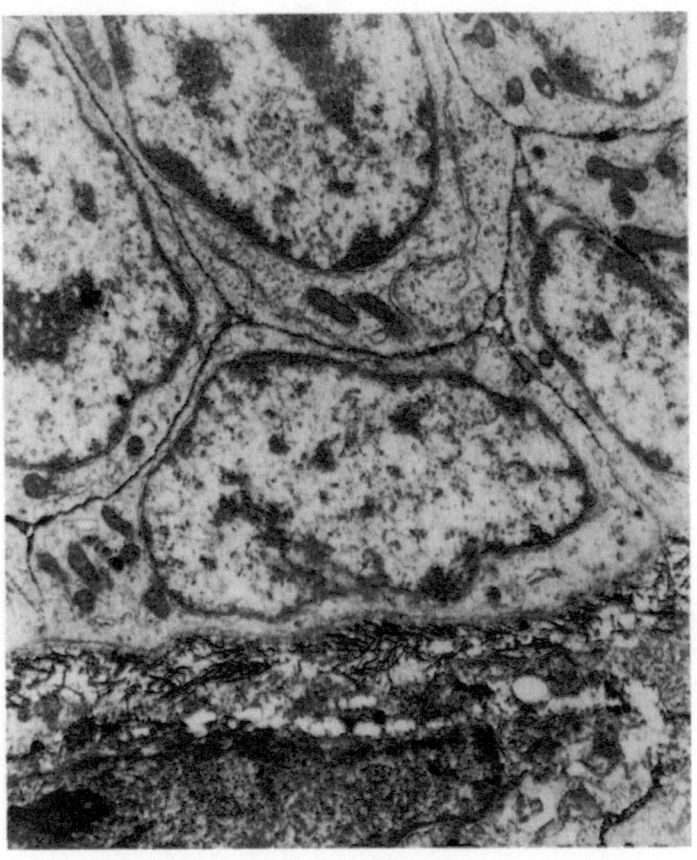

Fig. 5. Follicular wall of a secondary rat follicle.
Traces of lanthanum penetrate the intercellular
spaces. x 12.000

demonstrated firstly in the corpus luteum of pregnancy. Besides the well known effects on the pubic symphysis and on the uterus, an activity of the Rlx was detected also on the mammary glands, on the motility of spermatozoa in the seminal plasma and on the ovary. These effects are mainly dependent on the action that the relaxin exerts directly or indirectly on the connective tissue. As concerns the ovarian follicles, Rlx is supposed to act as a local hormo ne, that increases the release by hormone-primed granulosa

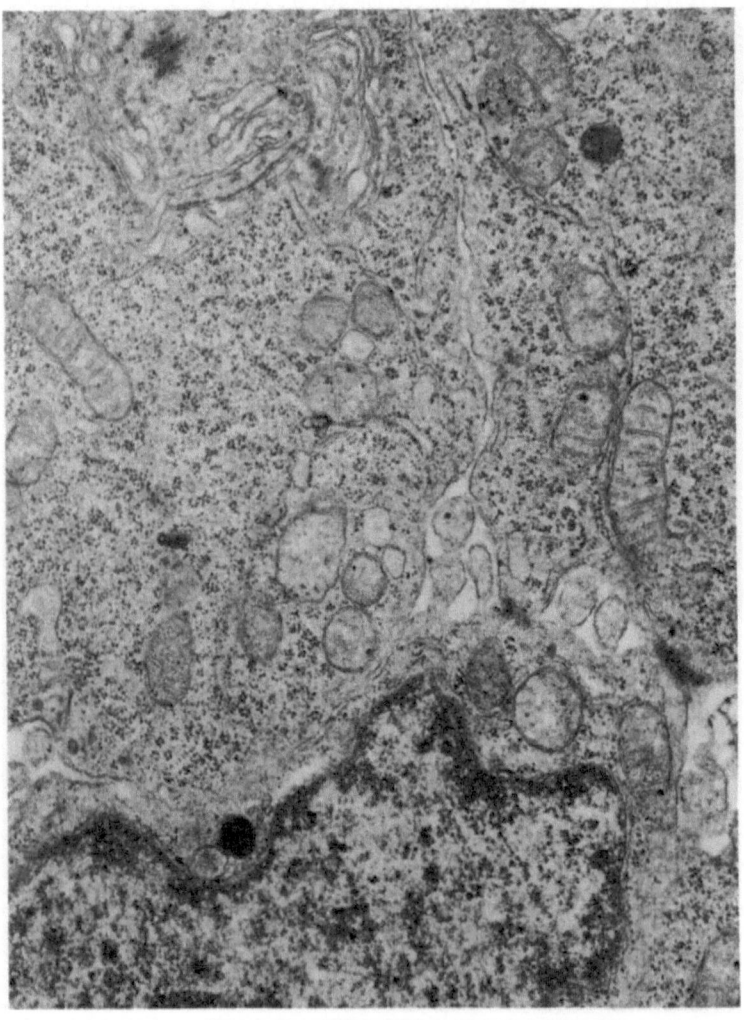

Fig. 6. Portions of granulosa cells of an evolving cavitary follicle. The cytoplasma contains a great
number of free ribosomes. x 20.000

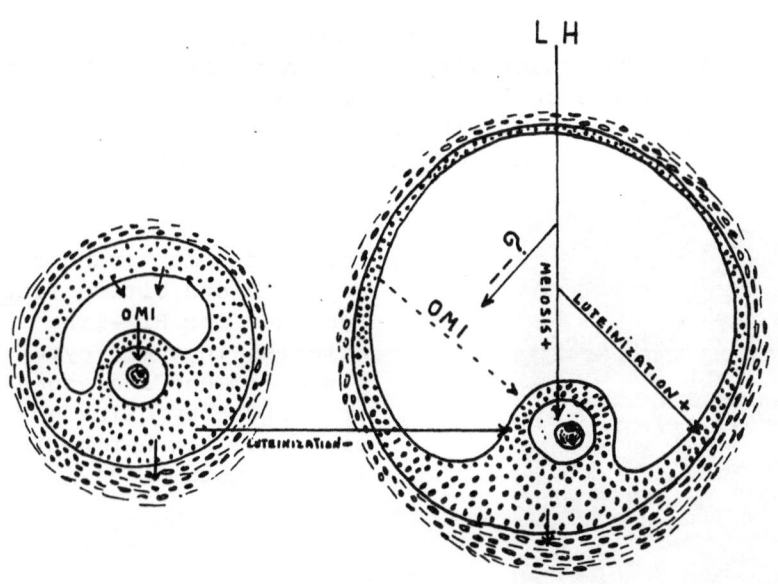

Fig. 7. Diagram showing how granulosa cells
produce an oocyte maturation inhibitory substance (OMI), that blocks
the oocyte in the prophase of the
first meiotic division, until the
LH surge. This substance is supposed to prevent also the luteinization of the granulosa cells.
LH favours the oocyte maturation
and probably suppresses the production and/or the activity of the OMI.

cells of collagenase, proteoglycanase and plasminogen acti
vator.
Recently, the production of Rlx by the wall of porcine ovarian follicles in vitro was demonstrated, as well the presence of this substance in the follicular fluid of human
cystic follicles.
    Our research (Balboni and Vannelli, 1983, 1986) with
the immunofluorescence and immunoperoxidase methods,using
the porcine anti-relaxin antiserum prepared by Steinetz,
demonstrated for the first time that immunoreactive Rlx is
present not only in isolated human granulosa cells, but
also in granulosa cells of the wall of human cavitary follicles in situ.
    The number of reactive cells progressively increases
with the follicular enlargement and maturation and subse-

quently during the luteal transformation. As concerns the
theca cells, the reactivity for relaxin is less evident
than in granulosa cells until the preovulatory stage, becoming quite similar in the luteinizing follicles. These
findings indicate that both granulosa and theca cells of

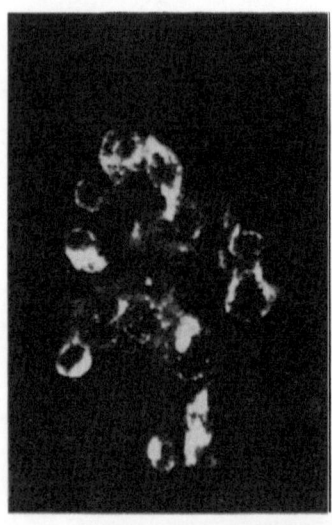

Fig. 8. Isolated human granulosa cells exhibiting fluo
rescence for Rlx-like
material in their cyto-
plasm. Immunofluorescen-
ce microscopy using por-
cine anti Rlx antibodies
(by Steinetz "R6")

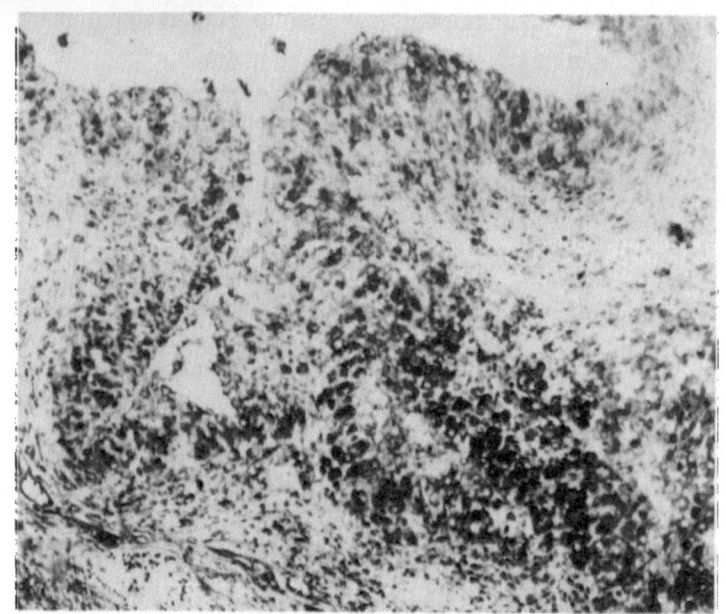

Fig. 9. Human luteinizing follicle. Intense positivity
for Rlx is shown by the granulosa cells.
Immunoperoxidase staining, after using porcine
anti-Rlx antibodies.

evolving follicles produce and/or store up relaxin. This fact fits well with the possible action of Rlx as local hormone involved in the follicular maturation and rupture.

Transferrin and transferrin receptor in human ovarian follicles (Figs.10-11)

It is well known that iron plays an important role in cell growth and metabolism. Transferrin (Tf) is a plasma protein devoted to the transport of iron. The specific transferrin receptor (Tfr) is a glycoprotein localized on the cell membrane of proliferating cells and is indispensable for the transport of Tf into the cell.

In previous research performed in collaboration with the Unit of Andrology of the University of Florence, we have demonstrated the presence of immunoreactive Transferrin in the Sertoli cells of human seminiferous tubules as well as the presence of Tf receptor in spermatocytes and early spermatids.

Tf and Tfr were not identified until now in the human ovary. Therefore we have studied the problem in this organ with the immunohistochemical methods. An antihuman Transferrin polyclonal antiserum and a murine monoclonal antibody to human transferrin receptor were used.

Our findings may be summarized as follows: 1) in the primary and secondary follicles the follicular cells are shar-

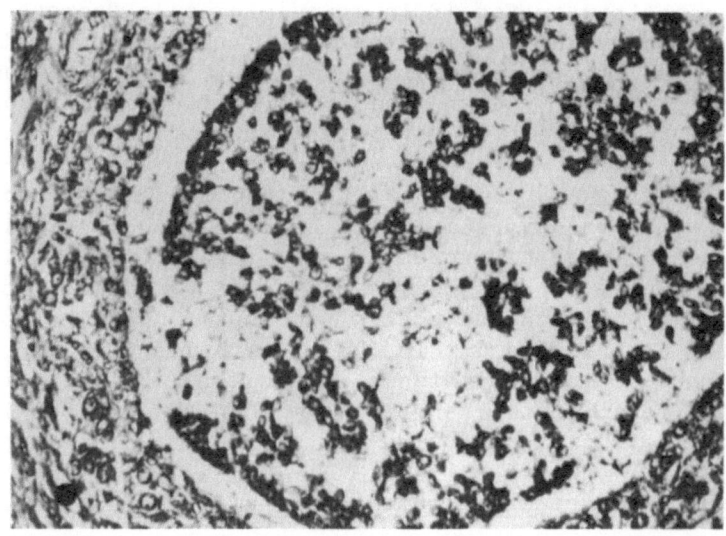

Fig. 10. Human ovary. Immunohistochemical demonstration of transferrin. The granulosa cells of a graafian follicle are intensely stained.●

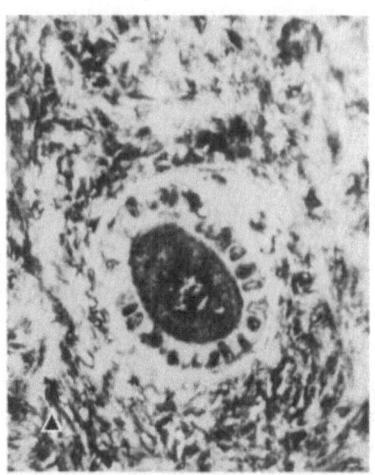

Fig. 11. Human primary follicle. Immunohistochemical demonstration of transferrin receptors. The oocyte is
strongly positive. ▲

ply reactive for both Tf and Tfr.The oocyte is intensely rea
ctive for Tfr; 2)in the cavitary follicles of small size, the
granulosa cells maintain their positivity for Tf and Tfr.
The oocyte presents an intense positivity for Tfr and a
feeble one for Tf. Some theca cells are moderately reactive
for both Tf and Tfr; 3) in the cavitary follicles of large
size, the positivity for Tf in granulosa cells decreases,
while quite constant remains that for Tfr. Oocyte is still
characterized by the positivity for Tfr. A number of theca
cells, but not all of them, are positive for both Tf and
Tfr; 4) in the involutive follicles only some theca cells
appear to be reactive for Tf and Tfr.

These findings suggest that Tf is a protein whose action
is utilized in the ovarian follicles. The granulosa cells
seem to be the site of production or storage of Tf, while
the oocyte,where the expression of Tfr is prominent, appears to be the main target cell of the substance. The positivity for both Tf and Tfr in the theca cells involved in
estrogen production indicates a possible rôle of Tf (of exo
gen origin) also in the regulating mechanisms of steroidoge
nesis.

Somatomedin C receptor in human ovarian follicles (Fig.12)

Somatomedin C or insulin-like growth factor I is a peptide growth hormone dependent that has a potent growth pro-
moting action on a variety of in vitro and in vivo cell systems. Evidence exists that somatomedins act in vivo by pa-

90

racine and/or autocrine mechanisms (see Underwood, D'Ercole, Clemmons and VanWyk, 1986).
Using a murine monoclonal antibody to human Somatomedin C receptor, we have investigated, in collaboration with the group of the Unit of Andrology of the University of Florence, the presence of Somatomedin C receptor in human ovarian follicles.

The preliminary results of this research have shown that the oocyte is strongly immunoreactive for SmC receptor. Some granulosa cells too present a sharp immunoreactivity. These data suggest that Somatomedin C is an important growth factor acting on the maturation process of the oocyte and on the proliferation of granulosa cells.

Follicular rupture and ovulation

The midcycle LH surge activates all the mechanisms involved in the follicular rupture and ovulation. A sequence of events occurs in the various components of the preovulatory follicle.
LH inhibits the follicular production of both estrogens and androgens and stimulates the process of luteinization of the granulosa cells. Furthermore, LH surge stimulates the production of Prostaglandis E and F, that influence the smooth muscle contractility and promote vascular permeability. In the same time, LH stimulates the synthesis of proteolytic enzymes acting in the disintegration of the apical wall

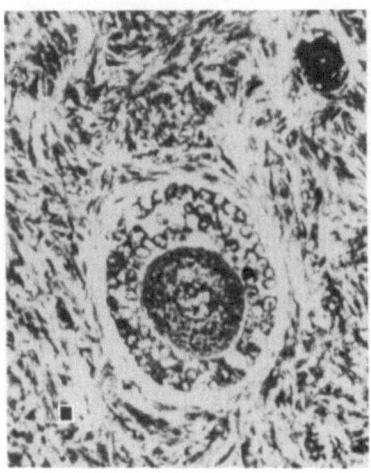

Fig. 12. Human secondary follicle. The immunohistochemical reaction of somatomedin receptors is positive in the oocyte and in some follicular cells.■

91

of the follicle.

The main morphological phenomena are represented by:
1) decrease in number of the linear gap junctions among
the granulosa cells, with a reduction of the cohesive for
ces that link these cells all around the follicle; 2) as
a consequence, the granulosa cells tend to assume an irre
gular outline with many superficial microvillosities, that
probably favour the development of hormone receptors and
of annular gap junctions; 3) development of the theca interna capillary network with interruption of the endothelial wall and of the capillary basal lamina, which favour
a leakage of fluid; 4) thinning and interruption of the
follicular basal membrane and disappearance of the so-called blood-follicular barrier; 5) production of lytic enzymes by the superficial epithelial cells and by the fibroblasts of the theca externa. An enzymatic digestion of col
lagen fibers occurs in the apical portion of the follicle;
6) contraction of the perifollicular smooth muscle cells
and squeezing of the follicle. All these mechanisms are
promoted by the LH surge, but probably they are to some
extent under the control of nervous influences.

Corpus luteum (Figs. 13-14-15-16)

Both the granulosa cells (granulosa lutein cells) and,
to a minor extent, the theca interna cells (theca lutein
cells) of the ruptured follicle contribute to the formation of the c.l.

The morphological sequences of c.l. formation and the
structure and ultrastructure of its cells are well known.
However, some facts can be pointed out.
The lutenization of the granulosa cells in the so-called
phase of transformation is characterized by the richness
of rough endoplasmic reticulum and free ribosomes in their
cytoplasm. The intense protein synthetic activity that oc
curs in this phase is probably related to the need of buil
ding up the cytoplasmatic structures (i.e. smooth endoplasmic reticulum) and the enzymes for the steroidogenesis.
The mitochondria may present both linear and vesicular
cristae.

A number of gap and septate junctions is present in
transforming and mature luteal cells, that facilitate the
interchanges of intercellular messengers such as cAMP.
The presence of microfilaments contributes to the process
of cellular reorganization in the c.l. and to the compar-
timentalization of the cytoplasm of luteal cells.
No substantial difference exists in histology, histochemi

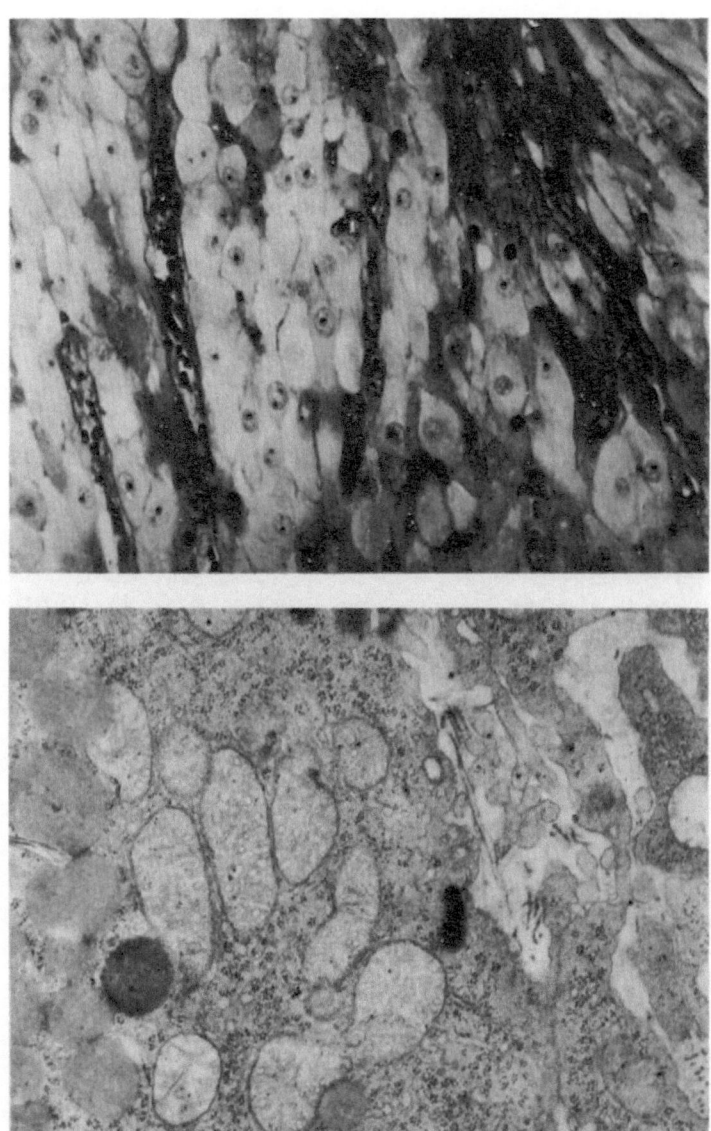

Fig.13. Human corpus luteum: lutein cells are supplied by a rich capillary network.

Fig.14. Lutein cells of a menstrual human c.l. in advanced stage of transformation. The cytoplasma contains abundant free ribosomes and some profiles of rough endoplasmic reticulum. Mitochondria show both linear and vesicular cristae. Cytóplasmatic processes project in the intercellular spaces. x 24.000.

stry and ultrastructure of menstrual and pregnancy corpora
lutea.

However, some peculiarities are shown by luteal cells
of pregnancy c.l. The smooth endoplasmic reticulum is more developed and forms folded membrane complexes and con-

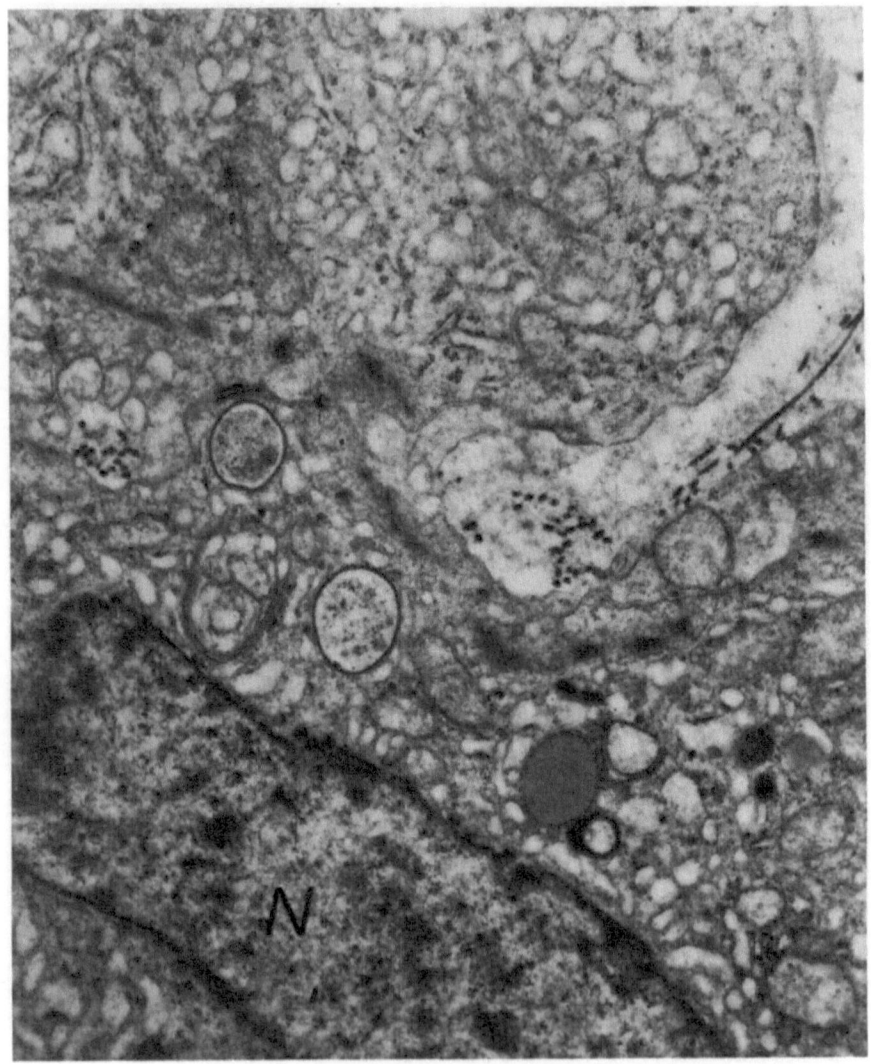

Fig. 15. Menstrual human corpus luteum.
In the cytoplasma of lutein cells a developed
smooth endoplasmic reticulum may be seen, as
well as some free ribosomes, lipid droplets,
and two annular gap-junctions. x 24.000.

centric membrane whorls.

The intercellular and intracellular system of channels is more prominent. Mitochondria are larger and pleomorphic. Numerous small granules of relaxin are closely associated to the cisternae of the rough endoplasmic reticulum.

Regressing corpus luteum (Figs. 17-18-19)

After the period of their activity both menstrual and pregnancy c.l. begin to regress as a consequence of hormonal mechanisms in which Prostaglandin F_2 is involved.

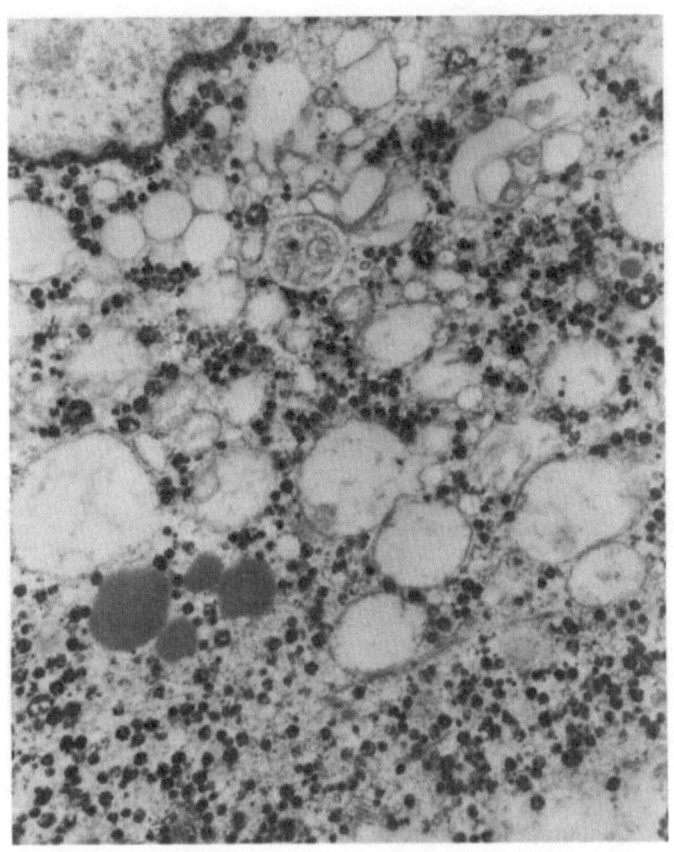

Fig. 16. Human corpus luteum of pregnancy.
In the cytoplasma of lutein cells an
electron dense material, steroid in
nature, may be detected with the
reaction of Friend and Brassil.

The luteolytic process is mainly characterized by degeneration of the luteal cells and by the proliferation of connective tissue which replaces the regressing tissue. A corpus fibrosus takes place.

The luteal cells accumulate in their cytoplasm lipidic material (triglycerides, phospholipids, cholesterol and/or

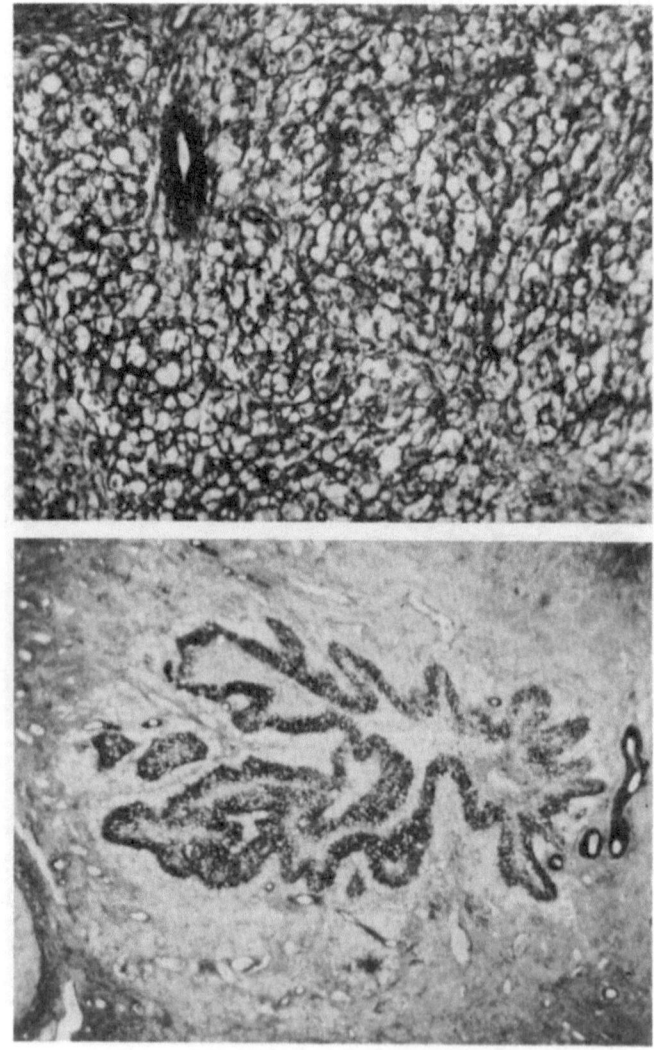

Fig. 17. Human ovary: regressing corpus luteum.
The degenerating lutein cells are replaced by the proliferation of the stromal connective tissue.

Fig. 18. Human ovary. Corpus fibrosus.

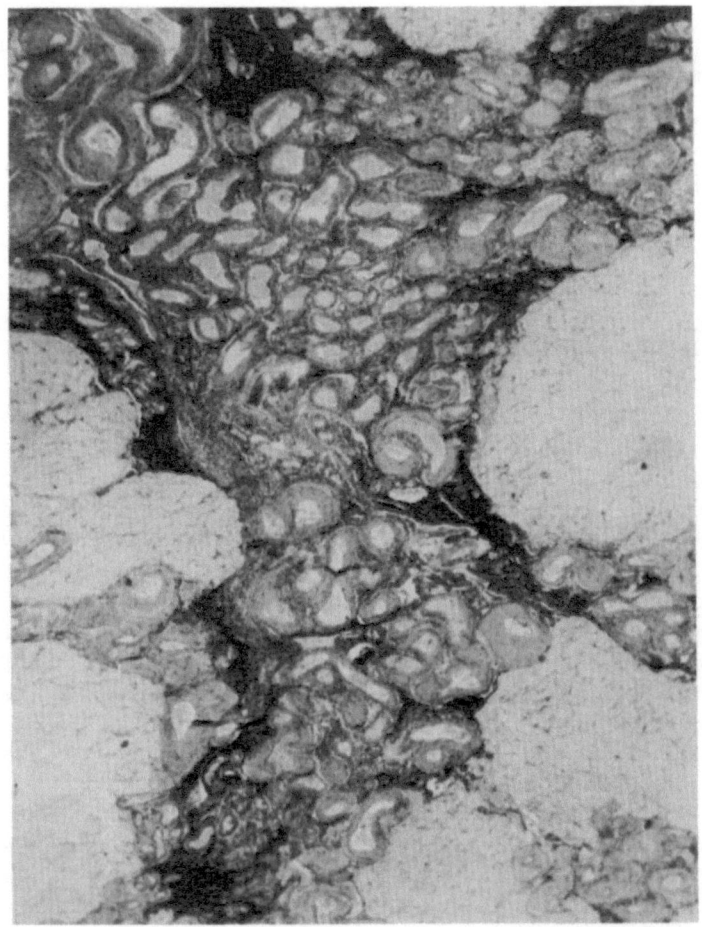

Fig. 19. Human ovary. Medullary zone, where the blood
vessels run a spiral course. Some corpora
albicantia can also be seen.

its esters). The positivity for the steroid dehydrogenase
decreases and enzymatic activities such as acid phosphatase and aminopeptidase become evident.
At the electron microscope, besides the presence of many
autophagic structures, an enlargement and fragmentation of
the smooth endoplasmic reticulum and a swelling of mitochondria may be observed.
At the end, the luteal cells degenerate completely and
only a patch of fibrous tissue remains. If the collagen
forming the corpus fibrosus undergoes hyalinization, a
corpus albicans takes rise.
Corpora fibrosa and albicantia, derived from the regres-

sion of corpora lutea and from the follicular atresia, represent a characteristic finding of the human adult and menopausal ovary.

References

1) Balboni G.C. Distribuzione del reticolo terminale adre nergico nell'ovaio. Boll. Soc.Ital.Biol.Sper.,1971,48, 84-86.
2) Balboni G.C. The problem of the ovarian stroma. Arch.Ital.Anat.Embriol.,1973,78,37-58.
3) Balboni G.C. and Teddę G. Observations electromicrosco piques sur les capillaires de l'ovaire humaine. Bull.Ass.Anat.,1973, 156,41-50.
4) Balboni G.C. Histology of the ovary. In: The Endocrine Function of the Human Ovary. edited by V.H.T.James, M. Serio and G.Giusti, Academic Press, London,1976,pp.1-24
5) Balboni G.C. Morphological remarks on the steroidogene sis in the human ovary. Folia Morphol.Czech.(Prague), 1977,25,46-52.
6) Balboni G.C. Morphological features of the human ovary during the menopause. In: The Menopause: Clinical. Endo crinological, and Pathophysiological Aspects, edited by P. Fioretti, L. Martini, G.B. Melis, and S.S.C. Yen, Academic Press, London, 1980, pp.191-200.
7) Balboni G.C. and Zecchi S. On the structural changes of granulosa cells cultured in vitro. Acta Anat., 1981,136-145.
8) Balboni G.C. et Vannelli G.B. La relaxine dans les fol licules de l'ovaire. Bull. Ass.Anat., 1983, 67,149-162.
9) Balboni G.C. and Vannelli G.B. Immunohistochemical de monstration of relaxin in the human evolutive and lutei nizing follicles. Verh.Anat.Ges.,1986,80,S.695-696.
10) Underwood L.E., D'Ercole A.J., Clemmons D.R., Van Wyk J.J. Paracrine functions of Somatomedins. Clin.Endocr.Metab.,1986,15,59-77.
11) Vannelli B.G., Orlando C., Barni T., Natali A., Serio M. Balboni G.C. Immunostaining of transferrin and transferrin receptor in human seminiferous tubules. Fertility and Sterility,1986,45,536-541.
12) Zecchi S., Repice F., Balboni G.C. On the granulosa cells of ovarian follicles. II. Identificaation of different morphological patterns of granulosa cells in evolutive follicles. Boll.Soc.Ital.Biol.Sper.,1981,57,583-587.

The application of electron microscopy in the evaluation of in vitro unfertilized human oocytes

G. FAMILIARI, S.A. NOTTOLA, S. PETRILLO, G. MICARA[1], C. ARAGONA [1], P.M. MOTTA

Department of Anatomy; [1]Department of Obstetrics and Gynaecology, University of Rome " La Sapienza", Rome, Italy

Introduction

In an "In vitro fertilization" program it is often difficult to judge the exact stage of maturation of an oocyte and its morphology due to the fact that the oocytes are always surrounded by cumulus cells at the time of collection. Other difficulties are due to the low resolution of the light microscope that is used in the oocyte evaluation. Therefore, the aims of this research are to study the fine morphology of "in vitro" unfertilized oocytes by means of scanning and transmission electron microscopy (SEM and TEM), and to compare ultrastructural data with light microscopical observations in order to have further useful morphological details for correctly evaluating "in vitro" mature oocytes.

Materials and Methods

Oocytes at different stages of maturation were obtained from women who had undergone pharmacological hormonal stimulation (clomiphene citrate + HMG-HCG). All the mature preovulatory oocytes (1), those that were mature at the time of sampling, and those oocytes that eventually reached maturity during "in vitro" incubation, were inseminated "in vitro" with washed spermatozoa that were previously incubated for one hour. 60 oocytes (which did not reach fertilization at 10-20 hours after insemination) were studied by SEM and TEM as well as by phase-contrast microscopy.

Oocytes were fixed in a solution of 3%. glutaraldehyde in cacodilate buffer 0.1 M at pH 7.4 plus culture medium without serum.

<u>TEM techniques:</u> after fixation for a few days, at 4°C, the oocytes

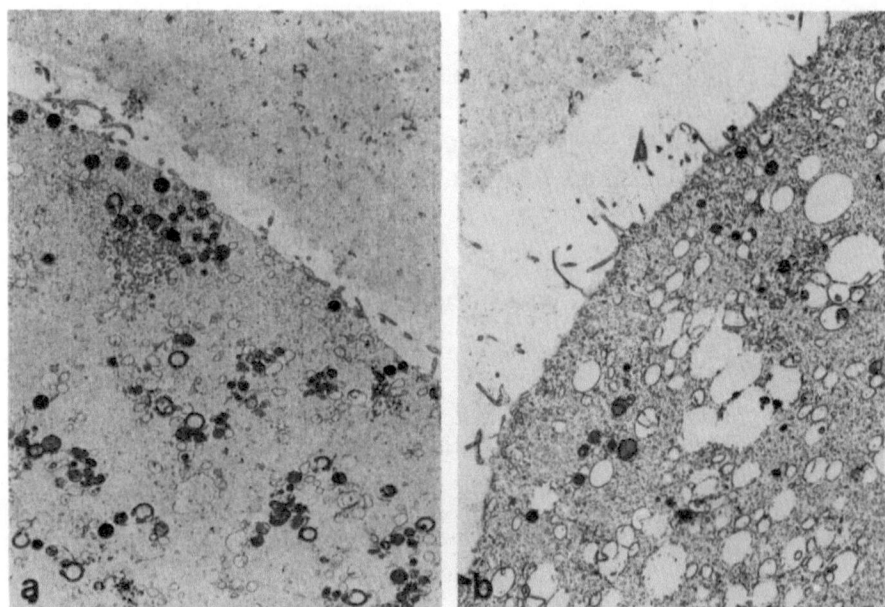

Fig. 1. Transmission electron microscopy.
a) Mature oocyte. Note the presence of cortical granules and
   mitochondria-smooth endoplasmic reticulum aggregates (4800 X).
b) Atretic oocyte. Note the extensive vacuolization throughout the
   cytoplasm (4800 X).

were washed in cacodilate buffer 0.1 M at pH 7.4, post-fixed in 1%
osmium tetroxide in cacodilate buffer, dehydrated in ethanol and
embedded in Epon 812. Sections, stained with uranyl acetate and lead
citrate, were observed on a Zeiss EM 9S2 electron microscope.

SEM techniques: after fixation for a few days, at 4°C, the oocytes
were washed in cacodilate buffer 0.1 M at pH 7.4, rinsed in distilled
water and placed in a steel micro-chamber whose base and top were 400-
mesh copper grids. They were then dehydrated in acetone, critical
point dried, mounted on the specimen holder and gold-coated. The
observations were made using a Cambridge Stereoscan 150 electron
microscope.

Observations

Preliminary results focused the ultrastructural characteristics of
the oocyte maturation as evidenced by TEM, and the surface details of
the zona pellucida as seen by SEM.

TEM: ultrastructural changes in mature oocytes consisted in an
increased number of cortical granules which apparently moved towards
the cell periphery, where they tended to form a single layer just

100

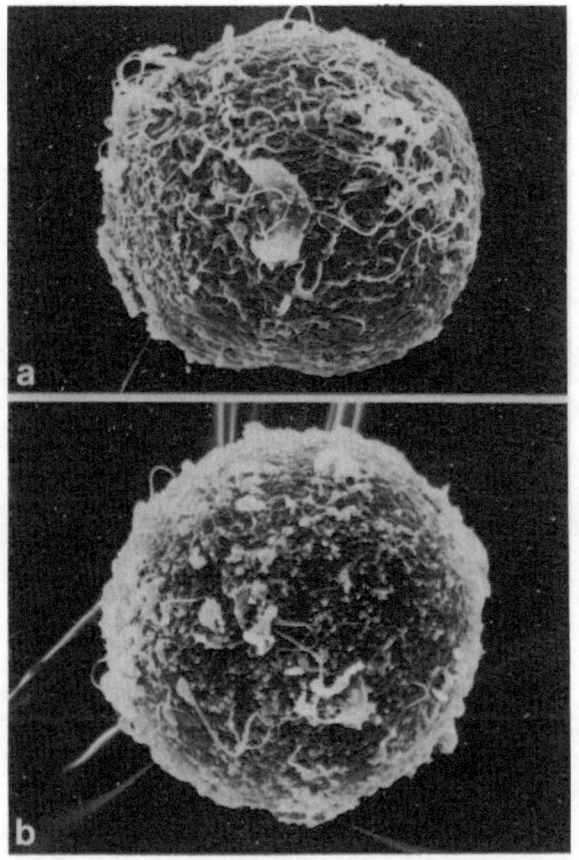

Fig. 2. Scanning electron microscopy.
a) Mature oocyte. The zona pellucida shows a sponge-like appearance
   (1500 X).
b) Atretic oocyte. The zona pellucida shows a more compact aspect
   (1500 X).

beneath the plasma membrane. There was a reduced amount of multivesicular bodies and vacuoles in the cytoplasm. The endoplasmic reticulum membranes were increased in number and the presence of mitochondria-smooth endoplasmic reticulum aggregates and mitochondria-vesicle complexes was noted (Fig. 1a).

Pathological features of unfertilized cultured oocytes included signs of delayed maturation, such as extensive vacuolization throughout the cytoplasm, decreasing in number and size or total absence of microvilli of the cell membrane and condensation of the ground substance (Fig. 1b).

<u>SEM:</u> the surface of the unfertilized cultured oocyte zona

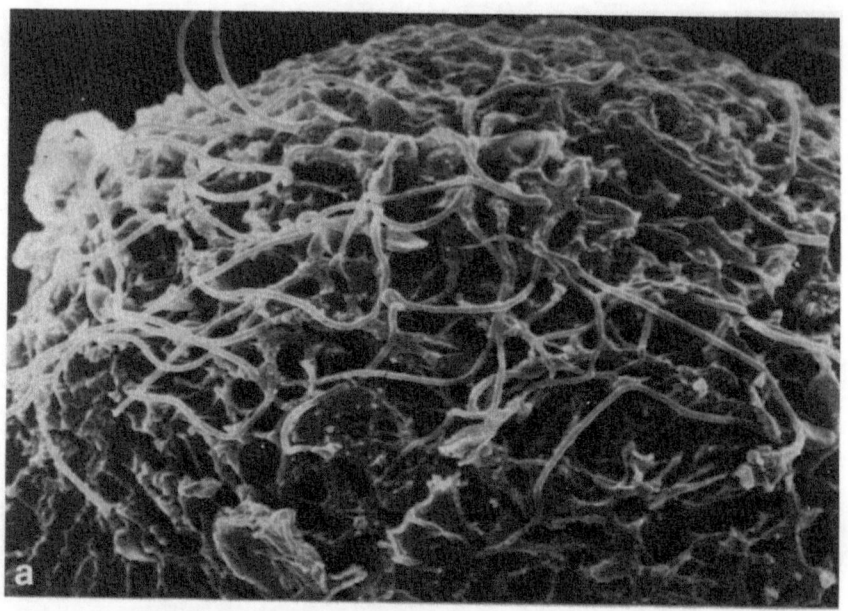

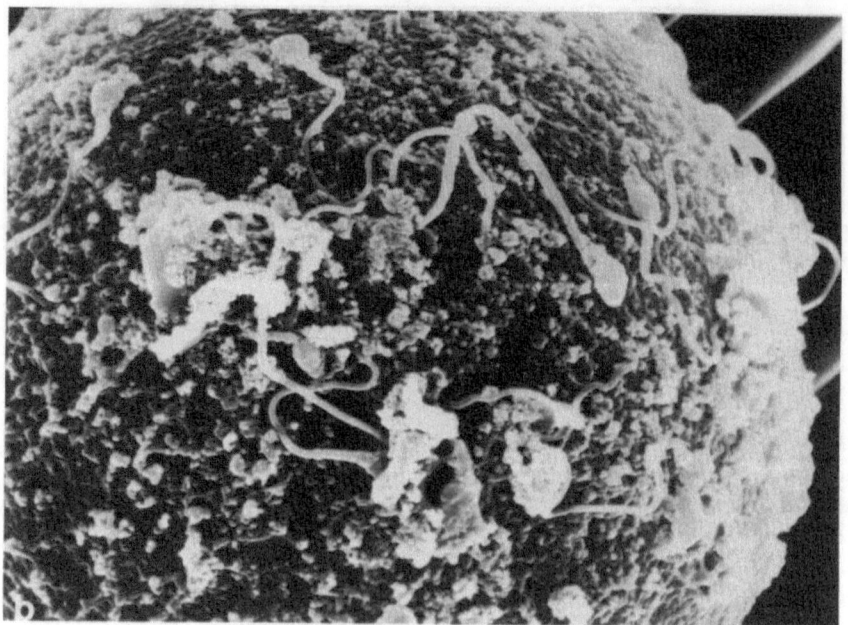

Fig. 3. Scanning electron microscopy.
a) Mature oocyte. Note the presence of numerous spermatozoa (4000 X).
b) Atretic oocyte. Note the presence of a few spermatozoa (4000 X).

pellucida was characterized by numerous fenestrations having a somewhat sponge-like appearance (Fig. 2a). Little changes in this structure covering mature or degenerated oocytes were observed. In fact, the zona pellucida of degenerating oocytes showed a more compact aspect (Fig. 2b).

Numerous spermatozoa were seen on the zona surface of mature oocytes, where penetration of the sperm head occurred almost tangentially to the surface of the zona (Fig. 3a). On the contrary few or no spermatozoa were seen in contact with altered zona pellucida (Fig. 3b) of atretic oocytes.

Discussion

TEM: oocyte maturation includes different steps: 1) breakdown of the germinal vesicle nuclear membrane, 2) first meiotic division resumption with formation of the first meiotic spindle, 3) extrusion of the first polar body, 4) formation of the second meiotic spindle, 5) arrest of meiosis at the second meiotic metaphase (1). When the oocyte reaches the second meiotic metaphase it is called "mature or preovulatory oocyte" (1).

"In vivo" oocytes acquire the capacity of maturation during their final stage of growth and follicular development. However oocytes, collected in the germinal vesicle stage, can also reach maturation during "in vitro" incubation (2).

The use of light microscopy is not always adequate for judging the exact degree of maturation: polar bodies can be mimicked by corona cells in the perivitelline space and pronuclei can be confused with large mitochondria-vesicle complexes in the ooplasm (3).

Our TEM observations showed in many unfertilized oocytes the presence of atretic features not detectable with the light microscope, probably related to failed fertilization.

SEM: SEM provided the most vivid and readily interpretable images of mammalian gametes and their interactions (4) when compared to light microscopy and TEM.

In SEM studies no changes were observed on the surface of the zona pellucida of mouse and hamster. This, in fact, appeared as a sponge-like structure 1) before fertilization, 2) at the time of fertilization, 3) during the development of the zigote and 4) following cleavage (5).

Only a few SEM micrographs of the interactions between human gametes have been previously presented (6-8). Sundstrom (7) showed also in the human the porous net-like appearance of the outer surface of the zona pellucida, but he further reported oocytes having a smoother or completely smooth surface of the zona. Therefore he concluded that a "smooth surface seems to be more common among non ovulatory oocytes, but no consistent differences in morphology due to the maturity of the oocytes can be found".

On the contrary, our observations show minor changes such as a more compact aspect of the zona pellucida of early degenerated cultured oocytes.

In addition, in the present study, a few spermatozoa are associated with the more compact zona of atretic oocytes, in contrast to the numerous sperm found on the sponge-like zona of mature oocytes. This is probably due to changes in surface properties of the zona.

Conclusions

The combined use of SEM and TEM compared to light microscopical observations seems to help a good deal to evaluating in detail the fine three-dimensional organization of in vitro unfertilized oocytes. These techniques also show a more consistent reconstruction of all the morphological events leading to "in vitro" fertilization.

Acknowledgements

The authors are most grateful to Mr. Antonio Familiari for the planning and construction of a steel micro-chamber for SEM preparations.

This work was supported by grants from CNR n° 85.00612.04 and MPI (40% and 60%).

References

1. Tsrafriri A, Bar-Ami S, Lindner HR. Control of the development of meiotic competence and of oocyte maturation in mammals. In: Fertilization of the Human Egg "in Vitro", eds. Beier HM and Lindner HR. Berlin, Springer Verlag, 1983:3-17.

2. Edwards RG. Conception in the human female. London, Academic Press, 1980.

3. Sundstrom P, Nilsson BO, Liedholm P, Larsson E. Ultrastructural characteristics of human oocytes fixed at follicular puncture or after culture. J Vitro Fert Embryo Transfer 1985:2;195-206.

4. Phillips DM, Shalgi R, Dekel N. Mammalian fertilization as seen with the scanning electron microscope. Am J Anat 1985:174;357-72.

5. Phillips DM, Shalgi R. Surface architecture of the mouse and hamster zona pellucida and oocyte. J Ultrastruct Res 1980:72;1-12.

6. Nilsson L. A portrait of the sperm. In: The Functional Anatomy of the Spermatozoon, ed. Afzelius BA. Oxford, Pergamon Press, 1975:79-82.

7. Sundstrom P. Interaction between spermatozoa and ovum "in vitro". In: Atlas of Human Reproduction by Scanning Electron Microscopy, eds. Hafez ESE, Kenemans P. Lancaster/Boston /The Hague, MTP Press Ltd, 1982:225-230.

8. Sundstrom P. Interaction between human gametes "in vitro": a scanning electron microscopic study. Arch Androl 1984:13;77-85.

Morphogenesis of the human foetal ovary: a morphometric approach

C. SFORZA, P. TARALLO[1]

Istituto di Anatomia Umana Normale, Milano; [1]Cattedra di Istologia ed Embriologia, Modena, Italy

INTRODUCTION

The morphogenesis of female gonads and the different stages of development, from the primordial germ cell to the mature oocyte, have been studied in several works, carried either on man or on other animals, mammals and not. The extragonadic derivation of oocytes is now well known, with their migration along the dorsal mesentery from the yolk sac to the genital ridge. The subsequent events are the tranformation of the primordial germinal cells in oogonads, the beginning of the prophase I of meiosis, and the organization of the gonad in germinal cords of somatic cells and gonocytes.

The next stage is the development of follicles, with the outgrowth of stromal cells from the medulla, and the separation of the clusters of gonocytes in primordial follicles.

In spite of the large amount of literature dealing with these problems, a morphometric approach is seldom carried out in the description of this phenomenon. The aim of this work is the construction of a morphometric model for the human foetal ovary, particularly concerned with the characteristics of developing follicles.

MATERIALS AND METHODS

We used 2 right ovaries, taken from 2 foetuses, aged 18 and 25 weeks, free from morphological and chromosomal anomalies. The ovaries were fixed in glutaraldeyde, postfixed in osmium tetroxide, embedded in epoxide and cut into 0.75 micron thick sections.

We considered as "primordial follicles" the ones in which one or two oocytes surrounded by a single layer of flat follicular cells were identifiable.

The following parameters have been evaluated, by using stereological methods:

- at lower magnification (37.5 X), the volume fraction of the ovarian cortex and medulla, and, in the cortex, the volume fractions of stroma and germinal epithelium;
- at higher magnification (730 X), the volume fraction of oocytes and of their nuclei, the size distribution and number of both of them, and, as a measure of the sinking level of the primordial follicles into the cortical layer, their percent distance from the medullary boundary.

RESULTS

Table I shows the volumes of the two studied ovaries, calculated both approaching the ovary with a triaxial ellipsoid (* $V = 4/3\,\pi\,abc$, where a, b, and c are the semiaxis of the ovary, i.e. half of the total length,, of the maximum diameter and of the minimum diameter of the transverse hilar section), and as the summ of products of the areas of the serial sections by the step (§). It also shows the volume fractions of cortex and medulla calculated by differential counts of points (°).

Table I. Volumes

Age weeks	* Ovarian volume mm^3	§ Ovarian volume mm^3	° V_v cortex	° V_v medulla
18	30.96	30.28	.918	.082
25	36.68	35.69	.883	.117

In order to study the spacial distribution of gonocytes in primordial follicles, we considered both different levels of distance from the equator of the gonad (hilar transverse section) and different areas in the same level. There are no differences between oocytes in the same level, but there is a highly significant difference between the considered levels, where the cells with bigger area stay in the levels nearer to the transverse hilar section of the ovary, and their distribution follows that of their nuclei.

In the 25-weeks ovary we particularly studied three levels of section at different distances from the transverse hilar plane, calculating the volume fractions (V_v) of ovarian medulla and cortex. In the cortex the parts belonging to the stroma, the primordial follicles and the clusters of gonocytes have been distinguished (Fig. 1).

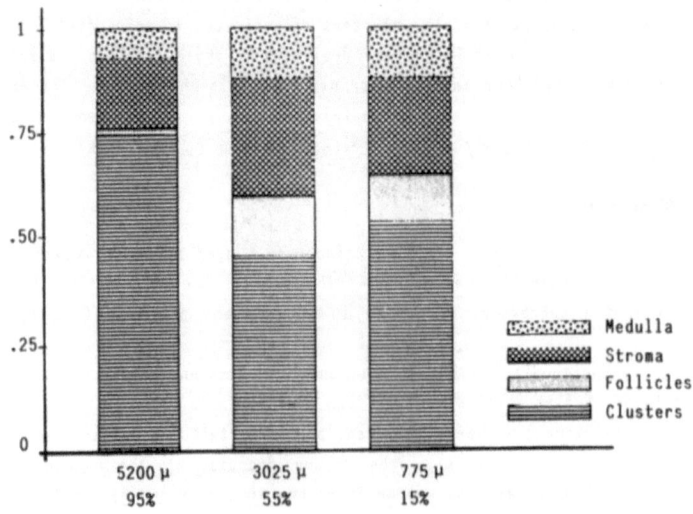

Figure 1. Volume fractions of ovarian medulla and cortex in 3 different sections in the 25-weeks ovary. In abscissa the distances of the respective sections from the transverse hilar plane, expressed in micron and as percentage.

106

Further, the percent distance of primordial follicles from the medullary boundary has been investigated as a measure of the sinking level of the follicles into the cortical layer (Fig. 2). While in the 18-weeks ovary the peak is about at half of the distance between the cortical/medullary boundary and the germinal epithelium, in the 25-weeks one the maximum crowding is nearer to the medulla. This concords with a development of primordial follicles from the cortex to the medulla, until they reach the stage of primary follicles. In fact, follicles surrounded by a layer of cubic epithelium (primary) could be seen only near the medullary boundary. This observation agrees with both the greater volume covered by follicles in sections next to the transverse hilar one (where the volume fraction of medulla is also bigger) and with a bigger volume fraction covered by the clusters (less mature oocytes) in polar sections.

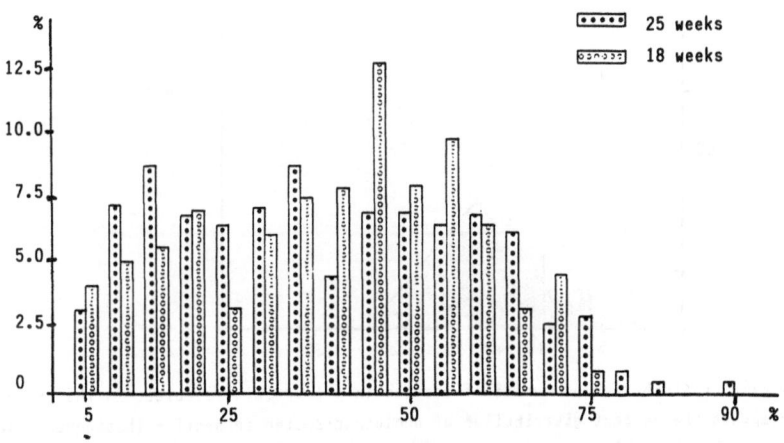

Figure 2. Histogram of the distance of primordial follicles from the ovarian medulla expressed as a percentage of cortical thickness.

Table II shows same morphometric parameters relating to the primordial follicles in the two ovaries (the cellular and nuclear areas have been measured in section; the cellular diameter has been calculated from mean area; the nuclear diameter and the number of follicles have been calculated from size distribution). Comparing the two ovaries it is easily seen that the maturation is greater in the 25-weeks ovary than in the 18-weeks one, with cellular and nuclear areas about one and a half bigger, but with like size distribution of nuclei (Fig. 3) and identical nucleus/cytoplasm ratio. The number of follicles in the volume unity, deduced from nuclear size distribution, is significantly higher in the older ovary.

Table II. Morphometric parameters

AGE	OOCYTES		NUCLEI		N/C	N_v follicles
weeks	Mean Area μ^2 (sd)	Mean Diameter μ	Mean Area μ^2 (sd)	Mean Diameter μ	Ratio (sd)	(mm^{-3})
18	433.04 (70.61)	23.48	98.05 (22.53)	13.47	.23 (.03)	6704
25	645.68 (76.80)	28.67	146.22 (28.27)	14.38	.22 (.03)	8007

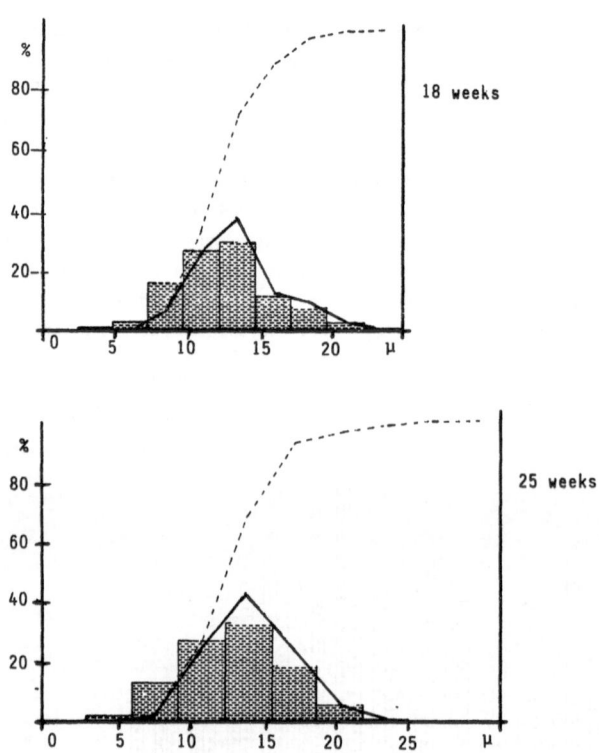

Figure 3. Size distribution of nuclei of oocytes in primordial follicles. In the figure are superimposed the percent distribution of nuclear diameter in section (histogram) the 3-D size distribution calculated with Schwartz-Saltykov method (frequency polygon in continued line) and the cumulative 3-D distribution (broken line).

CONCLUSIONS

The proposed morphometric model of foetal ovary could be used both for the study of a single moment of the formation of the gonad and for the comparision between subsequent stages of development. It could also be the starting point to compare normal ovaries and ovaries carrying different morphological and chromosomal anomalies.

Assessment of ovarian function by ultrasonography

A. IANNIRUBERTO

*Department of Obstetrics and Gynecology, City Hospital, Terlizzi,
Bari, Italy*

Pelvic ultrasound scans have assumed increasing importance in the monitoring the ovarian function and in the diagnosis and management of ovulatory disorders.

In 1972, Kratochwil demonstrated that pelvic echography not only showed the ovarian morphology but was also able to reveal the presence of follicular structures. Afterwards many Authors have used ultrasound techniques for follicular measurements, correlating the values of follicular growth with the ovulatory event.

Today, pelvic ultrasonography is widely used in various situations which can be schematically summarized:

1- Monitoring of follicular development in spontaneous and induced cycles.

2- Timing of the F.I.V.E.T. and G.I.F.T.

3- Prevention of the hyperstimulation syndrome and of multiple pregnancies.

4- Study of the follicular pathology.

5- Evaluation of the polycystic ovarian disease.

The ovary can be studied ecographically by using equipments with 3.5 MHz sector real time transducer. The patient is examined in a supine position, with a full-bladder technique, performing a series of longitudinal and transverse pelvic scans. To recognize the ovary, some landmarks can be taken in consideration, such as the ovarian vessels (superior pole) and the internal iliac artery (posterior pole).

Recently, vaginal intracavitary transducers have been employed with improved visualization of the follicles and easier pick up of the oocyte for the F.I.V.E.T.

The ovaries, that appear slightly echogenic and with clear limits, are measured in three planes (transverse, longitudinal and anteroposterior) in order to calculate the volume, assuming the shape to be that of prolate elipse using the following formula: Volume = D1 x D2 x D3 x 0.523.

Several other measurements have been carried out, according to the clinical and experimental needs, as such:

1- Ovarian size ($\frac{1}{2}$ x length x width x thickness).
2- Maximum surface area (maximum diameter x minimum diameter x 0.785).
3- Ovarian volume/Uterine volume ratio.

Important' is also the morphologic study of the ovary, including:

1- Ovarian echogenicity.
2- Presence or absence of ovarian cysts, their number, size and localization.
3- Ovarian margin, whether regular or irregular.
4- Ovarian capsule, whether thick or thin.

SPONTANEOUS CYCLE

In the 5th and 6th day of the menstrual cycle, in a normal ovary, some cystic structures of 5-6 mm. of diameter can be seen, that represent the follicles in the maturation process. After the 10th day, among all these follicles, only one increases progressively while the others disappear going in atresia(Fig.1,2,3). It has been demonstrated that follicles with dimensions inferior to 13 mm. are biologically unstable and can also evolve towards atresia. The daily growth has a linear course with an average increase of about 2 mm.

During the follicular growth in is possible to visualize a migration of the follicle towards the surface of the ovary, until it protrudes entirely from the ovary, appearing surrounded only by its own theca(Fig.4). In the period preceeding the ovulation, a close correlation is present between the dimensions of the follicle and the

110

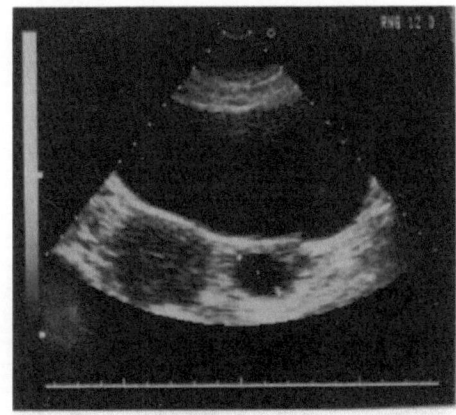

Fig. 1.
Follicle of 22 mm.
diameter.

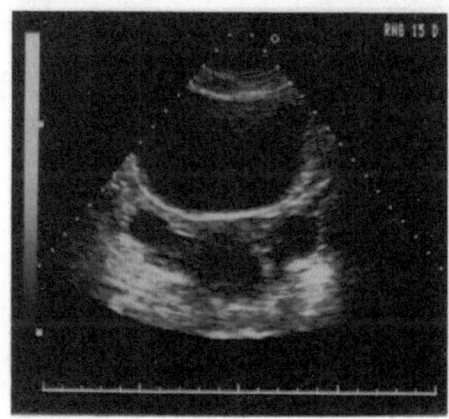

Fig. 2.
Follicles are seen in
both ovaries.

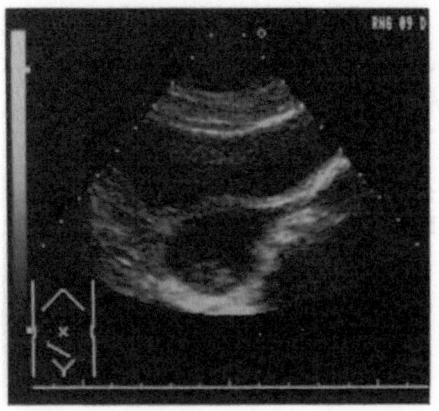

Fig. 3.
Large mature follicle.
It is possible to
visualize a few echoes
inside the follicle,
due probably to the
cumulus oophorus.

111

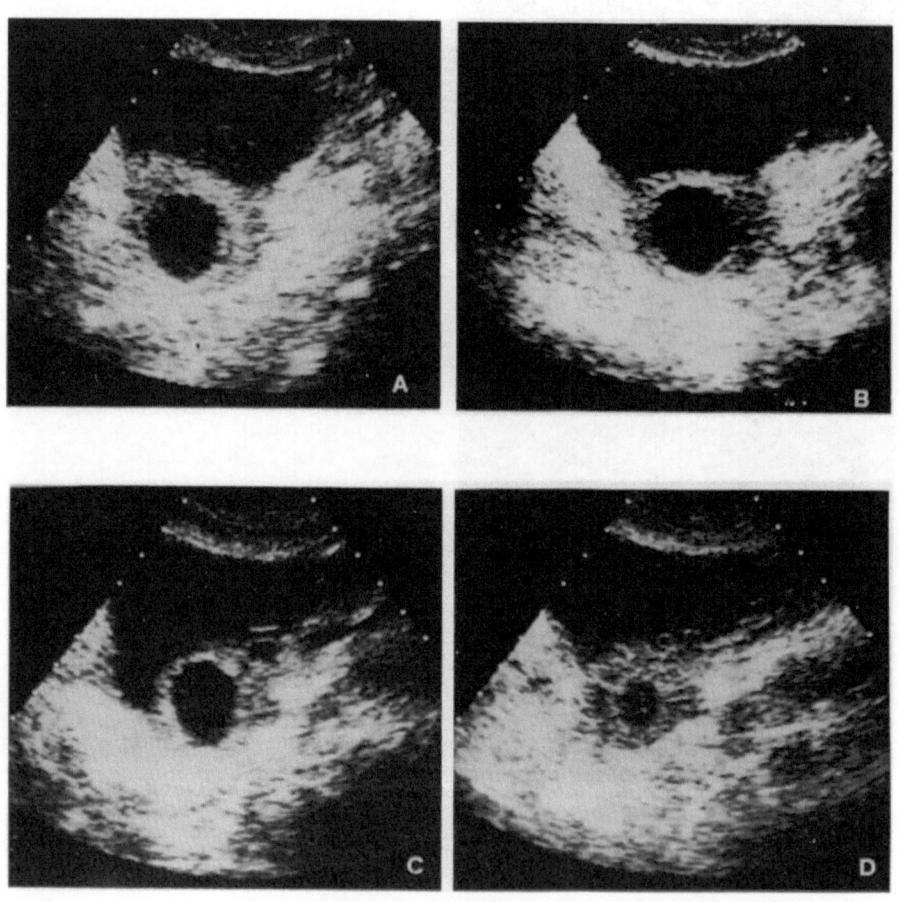

Fig. 4 (A–B–C–D). Follicular growth with migration of the
follicle towards the surface of the
ovary (A–B–C).

After ovulation, the follicle is
replaced by the corpus luteum (D).

17ß-estradiol plasma concentration. The association between the echographic finding and the E2 assay permits a more correct evaluation of the hormonal parameter, which can be correlated to the presence of only one normal follicle or of many follicles not yet matured.

By examining the follicular growth, the following steps can be established:

1- There is a short period in which the follicular growth is relatively rapid. This period takes place 48 hours before the ovulation simultaneously with the 17ß-estradiol peak.

2- There is a preovulatory plateau which coincides in about 80% of the cases with the LH peak level, sign of a reduction of the estrogenic secretion.

The ovulation usually occurs within 24 hours after the follicle has reached the maximum diameter.

The dimensions of the preovulatory follicle are still debated being included in a range that according to various Authors is between 15 and 26 mm.

According to Piclher, two ultrasonographic signs can be correlated with the imminent ovulation:

1- The appearance of a line of diminished echogenicity around the follicle, that appears 24 hours before ovulation.

2- The appearance of areas of discontinuity inside the follicular wall due to the progressive separation of the granulosa cells, that would appear 8 - 10 hours before ovulation.

The postovulatory events can be demonstrated by ultrasonography by taking into account the following patterns:

1- The preovulatory follicle disappears.

2- The follicle seems to be filled with discrete echoes.

3- The follicle collapses and is replaced within two to three days by the corpus luteum.

4- The appearance of an irregular cystic formation with dimensions inferior to those of the preovulatory follicle, referring to the corpus luteum (Fig.5).

5- The appearance of free cul-de-sac fluid (Fig.6).

This fluid collection could have been formed either from the follicular fluid or from a trasudate due to the ovarian congestion. This finding should be carefully valuated because sometimes the cul-de-sac fluid

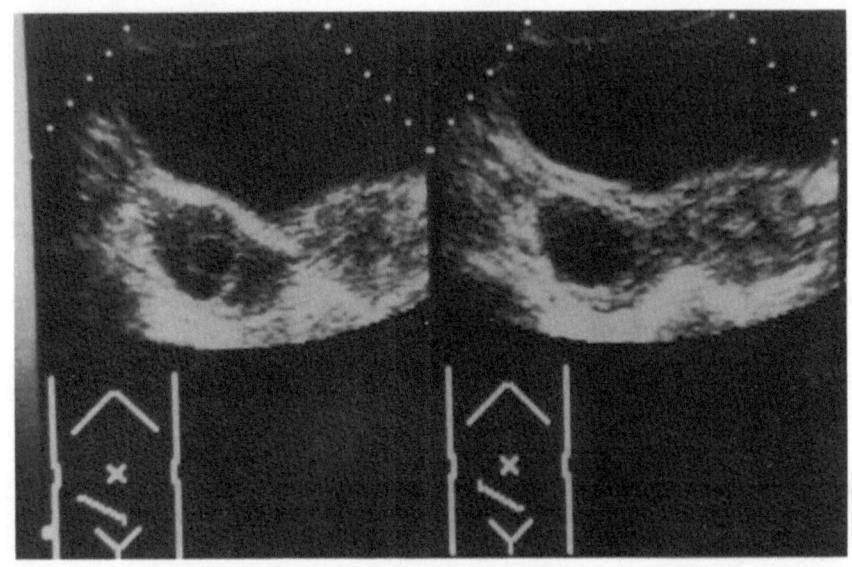

Fig. 5.   Corpus luteum. The follicle is replaced by a
cystic formation filled with discrete echoes,
with an irregular outline.

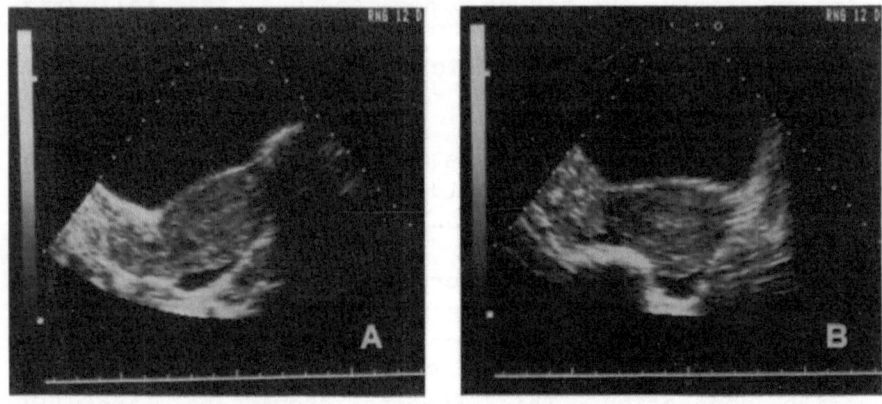

Fig. 6(A-B).   Appearance of free cul-de-sac fluid after
ovulation.

collection is to be attributed to the administration of liquids used for reaching an optimal filling of the bladder.

Finally, a short comment for the corpus luteum, which cannot be always seen because of the resolution power of the instrument. However two days after the ovulation it has a diameter of about 12 mm and persists for 4 to 6 days.

INDUCED CYCLES

In patients who are under pharmacological therapy for the induction of ovulation, it is possible to   reveal modifications of the ovarian morphology related to the number, the dimensions and the growth of the follicles. In fact, it is possible to visualize, bilaterally, small follicles in evolution and subsequently the outline of a dominant follicle.

Unlike the normal physiological cycles, it occ urs often to observe, in both the ipsolateral as well the controlateral ovary, competing follicles with a diameter greater than 10 mm. that do not reach the dimensions of the dominant follicle.

ultrasonography is useful   to determine the following points:

1- The achievement of the minimal follicular dimension for the most favorable condition for the induction of ovulation.

2- To   reveal   a possible multiple follicular development.

3- To permit an adequate pharmacological dosage for induction, either by increasing or by reducing the dosage amount.

4- To determine the timing of fertilization or of oocyte pick up.

The dimensions of the preovulary follicle are superior to those observed in normal cycles, being as follow:

Clomophene citrate: 22 ± 2 mm.
Gonadotropins   : 21 ± 2 mm.

In comparison to clomiphene, the gonodotropins induce more often multiple follicles in both ovaries.

The induction of the ovulation is carried out when the follicular diameter has reached about 20 mm. The detection of multiple mature follicles imposes to abstain from administering HCG, that represent the cause of an overstimulation. Some Authors have adopted an indicative score to prevent the overstimulation sy ndrome:

                                                      SCORE

1- Follicular growth
        1 follicle   1
        2 or more follicles   2
        slow   1
        rapid   2

2- Follicular diameter
        < 25 mm   1
        > 25 mm   2

3- 17ß-estradiol concentration
        < 400 µgr/ml   1
        > 600 µgr/ml   2

4- Previous clomiphene therapy   2

There is a good correlation between plasma concentration of E2 and the volume of all developping follicles. This correlation suggests that the plasma level of E2 represents the sum of the steroidogenic capacity of each follicle. As a consequence, in case there is only one follicle, the plasma concentration of E2 reflects the secretion and therefore the grade of maturity of that follicle. In presence of more follicles, ultrasonography is essential to interpret the values of E2, that are equal to those detected in the overstimulation syndrome, but in realty they refer only the presence of many follicles.
There can be large follicles with diameter greater than 20 mm., that do not produce a sufficient amount of E2 and that probably will be subjected to luteinization without rupture and, on the contrary, follicles of inferior dimensions that can give a good estrogenic secretion. This is the reason for which multifollicular cycles can be subjected to an overstimulation more frequently.

OVERSTIMULATION SYNDROME

The quantitative difference between the amount of gonadotropins needed to induce ovulation and those that can determine an overstimulation syndrome is very small. Therefore an overstimulation is an overhanging risk for every treatment. To prevent and maintain it within an acceptable risk it is important to evaluate echographically the basic conditions of every ovary. Infact it has been noticed that overstimulation is more frequent in the polycystic ovary.

The echographic findings of the overstimulation syndrome are different according to the degree of stimulation.

1- Light overstimulation.
It can take place after administering ciclophenile, clomiphene and HMG, with or without HCG·association. The ovaries have dimensions up to 50 mm. and present numerous cystic formations of different diameter, none of which usually go over the diameter of 30 mm. A modest free fluid collection in the Douglas can be noticed(Fig.7).

2- Moderate overstimulation.
The maximum diameter of the ovaries is less than 120 mm. There are follicles and corpus luteum formations enlarged and cystic, up to 50 mm. of diameter. It is usually present an evident free fluid collection in the Douglas, but there is no ascites.

3- Severe overstimulation.
It takes place soon after administering HCG. The picture is peculiar with the presence of acites in which the intestinal loops move. The ovaries are very large, exceeding 120 mm. in diameter and they are formed by a cluster of cystic formations of various dimensions (> 60 mm).

LUTEINIZED UNRUPTURED FOLLICLE SYNDROME (L.U.F. SYNDROME)

The ovulation process is made up by several distinct events, including ovum maturation and resumption of meiotic division, follicular luteinization and steroidogenic activity, follicular rupture and oocyte release. However, oocyte might become entrapped in a luteinized unruptured follicle (L.U.F.) with secretion of progesterone causing all the progestational changes without actual ovulation (Jewelewicz). Therefore, in these patients luteal phase plasma progesterone levels, basal body temperature and endometrial biopsy indicate that ovulation had occurred.

Although the ovulation events are controlled by independent mechanism, the trigger for each event seems to be the LH surge.

Inadequate surge of LH is responsible for the L.U.F. syndrome, but other factors may have an important role, like the oocyte maturation inhibitor, nonsteroidal luteinization inhibitors and stimulators, nonsteroidal gonadotropin regulatory substances, prostaglandins, proteolytic enzymes, cyclic adenosine monophosphate, transient psycological stress, attendant hyperprolactinemia and administration of clomiphene citrate.

Laparascopic evidence of the L.U.F. syndrome is based on the failure to demonstrate increased peritoneal fluid volumes and concentrations of progesterone and 17ß-estradiol that normally occur after follicular rupture and release of the follicular fluid into the peritoneal cavity.

The L.U.F. syndrome can be suspected by pelvic echography when it is found a loss of a clearly demarcated follicle wall and presence of intrafollicular echoes in the absence of a rapid decrease in follicular size or appearance of free cul-de-sac fluid. Usually, the dominant follicle undergoes a progressive increase in dimensions up to reach a diameter of 30-32 mm.(Fig.8).

These formations are called pseudocystic cysts, being different from the true luteinic cysts which appear 2 - 3 days after the mature follicle has collapsed.

118

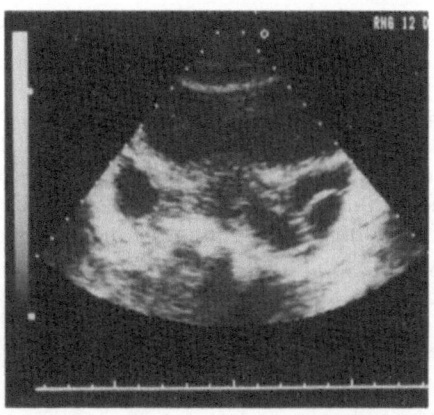

Fig. 7.   Light overstimulation after clomiphene therapy.
The ovaries have numerous cystic formations of
different diameter.

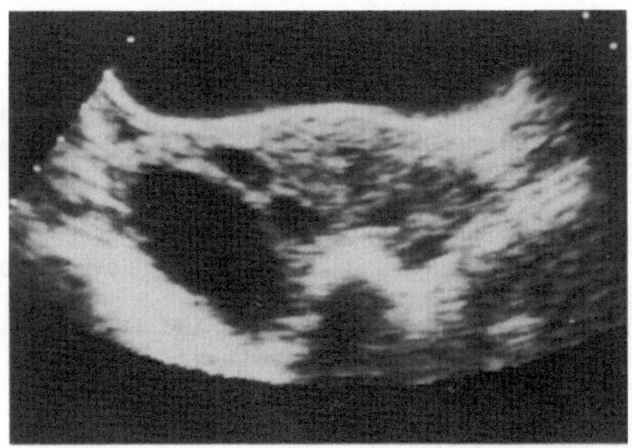

Fig. 8.   Luteinized unruptured follicle syndrome (L.U.F.
syndrome). The dominant follicle undergoes a
progressive increase in dimensions without
rupture. The follicular wall is not clearly
demarcated. With the ultrasound gain control it
is possible to visualize fine echoes inside the
cystic formation.

THE POLYCYSTIC OVARY (PCO) AND MULTIFOLLICULAR OVARY(MFO)

The polycystic ovary (PCO) is a syndrome referred to a group of conditions, whose common characteristic is the presence of multicystic ovaries which secrete a high quantity of androgens. the multicystic ovary is a sign and not a diagnosis, therefore the echographic examination by itself can only give a diagnostic suspicion which is to be confirmed by the hormone measurements.

According to ultrasonographic data and the hormone measurements, it is possible to make a distinction between women that have polycystic ovaries and women who instead present only multifollicular ovaries (MFO).

MULTIFOLLICULAR OVARY (by J.Adams and Coll. The Lancet, December 21/28, 1985)

1- Normal ovarian volume.
2- Decreased uterine area.
3- Ovaries filled with six to ten cysts 4-10 mm. in diameter.
4- Absence of stromal hyperplasia.
5- Absence of hirsutism.
6- Amenorrhea or oligomenorrhea.
7- Serum LH concentrations are similar to those in normal subjects during the mid-follicular phase (days 5-9) of the menstrual cycle.
8- Low FSH.
9- History of weight loss.
10- Good response to induction of ovulation with pulsatile LHRH treatment.
11- Growth of the dominant follicle and regression of the remaining cysts with LHRH treatment.

POLYCYSTIC OVARY (PCO)

1- Ovaries usually enlarged (Fig.9).
2- Increased ovarian volume/uterine volume ratio.
3- Multiple cysts (ten or more) 2-18 mm. in diameter distributed around the ovarian periphery. Less commonly multiple small cysts 2-4 mm. in diameter distributed in the abundant stroma (Fig.10).

120

4- Increase in stromal tissue.
5- Oligomenorrhea, amenorrhea, regular cycles or dysfunctional uterine bleeding.
6- Hirsutism.
7- Elevated serum LH and testosterone.
8- Persistent polycystic pattern with LHRH treatment.

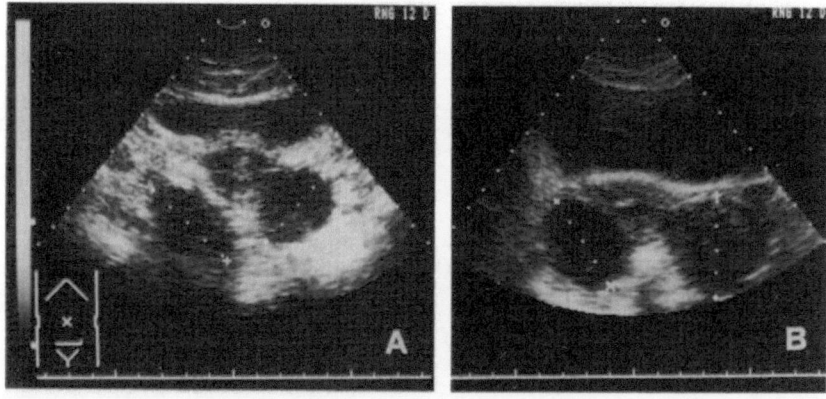

Fig. 9 (A–B).   Polycystic ovarian syndrome. The ovaries are enlarged, with a hyperechogenic capsula.

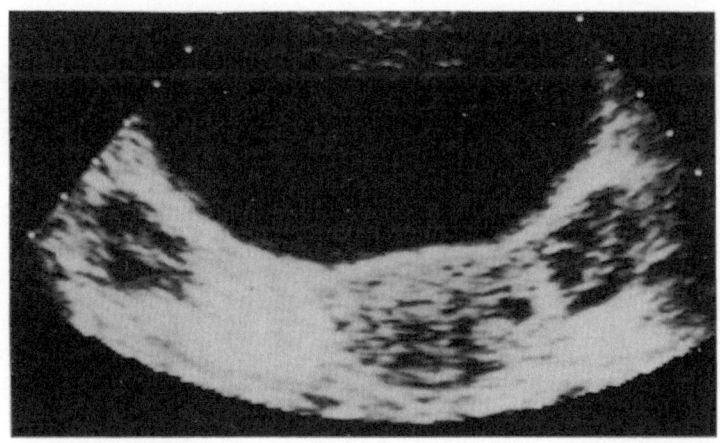

Fig.10.   Polycystic ovary. Multiple small cystic formations are distributed in the ovarian stroma. The capsula is hyperechogenic.

In conclusion, it is emphasized the great importance of pelvic echography for the diagnosis of ovulatory disorders. However, ultrasound studies alone are unable to establish a correct diagnosis due to the variability of ovarian morphology in many pathologic conditions.

The difficulty probably will be overcome in the near future with the improvement of the ultrasonic instruments and with new scanning approaches.

Morphological basis of female orgasm

R. STEFANI

*Institute of Comparative Anatomy, University of Cagliari,
Cagliary, Italy*

INTRODUCTION

Mammalian females who ovulate only after orgastic reflex during coitus are called reflex ovulators.Spontaneous ovulators ovulate independently of orgasm.Man is one of them.I think that the first condition is more primitive than the second.

In the first group,LH-release is mediated by the hypothalamus and a cholinergic factor is implicated in the control of LH action of the pituitary.According to ARON,ASCH and ROSS there is no sharp borderline between reflex and spontaneous ovulators(1).

These statements are sufficient to demonstrate the role of the orgastic reflex in mammals.In man,almost nothing is known about its implication in the physiology of the ovary and genital apparatus.

Now,there is no anatomical doubt that during human coitus the penis does not come in direct contact with the clitoris.The clitoris normally fails to be stimulated by penile thrust.So that in a human sexual intercourse the penile intromission is very often insufficient to release orgatic reflex in the woman.

Coital frigidity thus represents a very special instance of situational orgastic difficulty.According to KAPLAN,there are millions of women who are sexually responsive and often multiply orgastic,but who cannot

123

have an orgasm during intercourse unless they receive simultaneous clitoral stimulation.Coitus in itself is not sufficiently stimulating to enable them to reach a climax.These are the women,HELEN KAPLAN says,who pose a dilemma for the clinician.Are they neurotic ? Can their complaint be attributed to the fact that their husbands are ineffective lovers ? Are there deep-seated disturbances in the marital relationship ? Or is it a failure to reach a climax during intercourse a normal variation of female sexuality ? (2).

THE THEORY OF VENTRALIZATION OF CLITORIS

In the primitive anatomical condition by pronograde Primates,the vagina,as in other quadrupeds,is a funnel-like canal divided into a uterine vagina and vestibule.Urethral orifice and clitoris belong to the ventral floor of the vestibule and are of course not externally visible.

The penis intromission does necessarily bring contact with the clitoris glans.Orgastic reflex arouses automatically and there is no use to distinguish a vaginal from a clitoral orgasm.

During the process of evolution of the human upright posture a formation of the muscular pelvic floor took place,derived from the primitive ischio-caudal,pubo-caudal and ilio-caudal muscles.

Quadrupeds have an ischio-pubic symphysis which involves the whole ischio-pubic ramus.In clinograde apes the ischial tuberosities do not meet and there is a beginning of a subpubic space which becomes a marked feature in the human pelvis.From the most generalized ischio-pubic symphysis of quadrupeds we pass gradually to the short and specialized pubic symphysis of man (3).

Now,during the evolution of the upright posture,the clitoris and the urethra changed their position with a process of ventralization of these organs.The clitoris,closely bound to the inferior margin of the symphysis followed the migration of the rim in a ventral direction.The whole vestibule has been drawn up and become external to the vagina as we see in the human female.Details and figures are given in a previous paper (4)

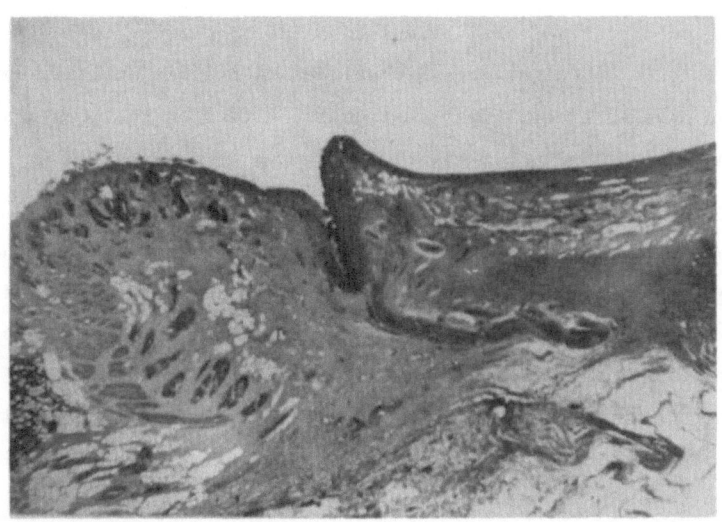

Fig.1.Longitudinal section of the vagina floor of the cat with the
intravaginal clitoris.

DISCUSSION

KRANTZ in a microscopic study of the innervation of the human vagina(5)
has shown that no nerve corpuscles are present in the muscularis and mu-
cosa.Just occasionaly,free endings were observed in the mucosa and tunica
propria.The vagina wall,as every gynecologist knows,can be handled without
any particular sexual response.Quite different is the response of clitoris
where masses of corpuscles of Meissner,Merkel,Pacini,Ruffini and Dogiel-
Krause are concentrated.

Freudian distinction between vaginal and clitoral orgasm is no more
valid and must be definitly radiated.Orgasm is a physiological reflex
that implies as a response the whole genital complex.

Coital frigidity is a myth.HELEN KAPLAN says:unless severe physical
or mental illness exists,or there is insurmountable marital discord,all
women seem to be capable of responding sexually and of having orgasm.The
inability to achieve orgasm on coitus does not necessarily mean that the
woman is neurotic;nor does it mean that she is sexually inadequate,or imply any failure on the husband's part.Millionsof couples have developed
an excellent and mutually gratifying sexual relationship although the

125

wife depends on direct clitoral stimulation for orgastic release.The obsessive insistence on coital orgasm,concludes KAPLAN,when this is not feasible can have an exceedlingly destructive effect on the couple's sexual relationship.

CONCLUSION

The comparative anatomy has shown that during the evolution of the man's upright posture,the pelvis underwent a gradual modification of structure that brought to an opening of the pubic symphysis and consequently to a ventralization of the vestibule with exteriorization of clitoris which became external to vagina.The enlargment of the pelvic outlet was necessary for parturition,but copulation occurred mainly without female orgasm and this phenomenon played a colossal role in the sexual history of   mankind.

REFERENCES

1. ARON C.,ASCH G.,ROOS J. Triggering of Ovulation by Coitus in the Rat.Int.Rev.Cytol.,1966:20;139-172.

2. KAPLAN H.S. The new Sex Therapy.Brunner/Mazel.New York,1974.

3. STEFANI R. Some consequences of the evolution of the human pelvis. Arch.It.Anat.Embr.1984:89;131-140

4. STEFANI R. Sull'evoluzione del clitoride umano.XL Conv.Soc.It.Anat. Milano-Como,25 sett.1984.Suppl.38;431-432.

5. KRANTZ K.E. Innervation of the human vulva and vagina.Obst.Gynec.1958: 12;382-396.

Observation of the ovary through endoscopy: from the morphology to the physiopathology of an organ

N. GARCEA, R. DARGENIO, S. CAMPO, P. SICCARDI

Clinica Ostetrica e Ginecologica (Dir. Prof. S. Mancuso), Università Cattolica del Sacro Cuore, Rome, Italy

INTRODUCTION

The use of endoscopy in gynaecological diagnosis is today universal. Laparoscopy (1,2,3) and, more recently, hysteroscopy (4,5) have amply demonstrated their high degree of diagnostic reliability and their respective indications have been minutely described. Moreover, when used in infertile patients, these techniques not only permit a more meticulous examination, but make it possible to carry out highly important therapies, such as techniques for assisted fecundation.

Laparoscopy has thus become an almost inevitable step in the case history of the infertile couple, placing the gynaecologist who carried it out under the moral obligation to gather as much information as possible from this technique which, though relatively simple, is nonetheless invasive.

Seen in this light, observation of the ovarian morphology, an often loquacious indication of the endocrine situation, assumes considerable importance in the diagnostic definition of patients who are still, in a third of cases, labelled as affected with unknown fertility.

OVARIAN MORPHOLOGY IN PHYSIOLOGICAL CONDITIONS

It is known that ovarian morphology changes in accordance with the menstrual cycle: thus laparoscopy in a woman who ovulates will show the different aspects of ovulation. If carried out prior to ovulation it will show the follicle in the different phases of growth and maturation that culminate in ovulation. If, on the other hand - as is normally the case - it is carried out in the post-ovulatory phase, it will be possible to verify

127

the presence of the corpus luteum and, if the ovary is in the early luteal phase, see the stigma. A LUF syndrome may thus be excluded on this basis.

Lastly, the presence of scars from previous ovulations will permit retrospective evaluation of the ovarian function.

OVARIAN MORPHOLOGY IN FUNCTIONAL PATHOLOGY

When laparoscopy is carried out in women affected with chronic anovulation it can reveal primary or secondary pathological conditions of the ovary. Among the most common are:

a) Ovarian hypoplasia, which may be mild, moderate, severe or streak-like. These conditions are generally due to some pathology of the tonic secretion of gonadotropins or, more frequently in the case of streak-like ovaries, marked by primary ovarian deficiency.

b) The polycystic ovary and the Stein-Leventhal type of ovary, signalling damage to the cyclic secretion of gonadotropins.

c) The smooth ovary, bearing no signs of ovulation, which probably represents a process of development towards endocrine conditions such as ovarian polycystosis.

d) The simple ovarian cyst

e) The cerebroid, menopausal ovary, no longer able to respond to the pituitary message.

LAPAROSCOPY IN THE DIAGNOSIS OF MENSTRUAL IRREGULARITIES

On the basis of the above observations we wondered how diagnostically reliable is endoscopic observation of the ovary in regulation to the most frequently found menstrual irregularities. For this reason we carried out a retrospective analysis of the case histories of 1078 women in whom laparoscopy had been performed for this purpose: we found ovarian pathologies in 78% of cases and "normal" ovaries in the remaining 22% (Table 1).

128

TABLE 1 Laparoscopic reports in patients with menstrual irregularities

Laparoscopic report	no. of patients	%
Ovarian pathology	848	78
Other pathologies	104 ⟩ 230	10 ⟩ 22
Normal report	126	12

TABLE II Laparoscopic reports (in%) distributed according to symptoms

Laparoscopic Report	Primary Amenorrhea	Secondary Amenorrhea	Hypo oligomen.	Metror- rhagia	Hyper- polymenor.
Normal	0.9	7.8	12.1	24.2	19.8
Ovarian hypoplasia	50.0	32.7	15.6	7.8	10.4
Streak-like ovary	11.2	1.7	--	--	--
P.C.O.	3.8	25.0	46.6	13.8	12.9
Smooth Ovary	10.0	23.2	12.6	6.0	17.2
Ovarian cyst	--	0.9	6.0	20.7	29.3
Cerebroid ovary	--	7.8	3.4	3.4	5.2
Malform.	20.7	0.9	--	0.9	--
Other	3.4	--	3.4	23.2	5.2
Total	100.0	100.0	100.0	100.0	100.0

When the individual symptomatic situations were examined in detail (Table11) it was noted that different degrees of ovarian hypoplasia were found in over 60% of cases of primary amenorrhea. In the case of secondary amenorrhea the findings were evenly distributed between hypoplasia, smooth ovary and polycystic ovary. In about 60% of cases of hypo-oligomenorrhea the existence of polycystic or smooth ovaries was revealed. In metrorrhagia and polymenorrhea, while the most frequent findings were normal ovaries and simple ovarian cysts, the laparoscopic reports were fairly evenly distributed.

It would seem from the above data that there is a fairly marked correlation between laparoscopic findings and sympotomatological situations peculiar to certain endocrine conditions such as hypopituitarism associated with primary amenorrhea and PCOS associated with hypo-oligomenorrhea-type menstrual alterations. This analysis was, on the other hand, less precise in cases of laparoscopic reports associated with less pathognomonic symptomatological situations such as secondary amenorrhea, metrorrhagia and hyper-polymenorrhea.

LAPAROSCOPY IN THE DIAGNOSIS OF DEFINED ENDOCRINE CONDITIONS

In order to further to verify the results of the previous analysis, we examined in perspective the laparoscopic reports found in well defined endocrine situations such as secondary hypogonadotropic hypopituitarism and PCOS.

In hypopituitarism associated with primary amenorrhea we found varying degrees of ovarian hypoplasis in 84% of patients against 16% of other findings (Table III).

If the analysis is carried out on patients selected on the basis of gonadotropin levels constantly below 10 m.I.U./ml, we find laparoscopic reports of ovarian hypoplasia in only 40% of cases. Parallely, the hormone levels of the patients with ovarian hypoplasia observed during laparoscopy are systematically lower than 10 m.I.U./ml in 42% of cases. (Table IV)

It should nevertheless be noted that in the remaining 58% of cases the hormone levels are systematically higher than 50 m.I.U./ml pointing to ovarian receptor insufficiency. This is confirmed if we separate the findings of ovarian

TABLE III Laparoscopic reports of patients with primary and secondary amenorrhea due to hypopituitarism.

Laparoscopic report	no. of patients	%
Ovarian hypoplasia	39	80
mild 7		
moderate 15	41	84
severe 17		
Streak-like ovary	2	4
Smooth ovary	3	6
Polycystic ovary	1 ⌉ 8	2 ⌉ 16
Other	4	8
Total	49	100

TABLE IV Relationship between laparoscopic findings of ovarian hypoplasia and low levels of gonadotropins.

100%	Ovarian hypoplasia or streak-like ovary	LH and FSH <10 mIU/ml
LH and FSH <10 mIU/ml	40% (vs 60% other laparoscopic report)	– –
Ovarian hypoplasia or streak-like ovary	– –	42% (vs 58%: LH and FSH >50 mIU/ml)

hypoplasia considered globally from those of streak-like ovary and compare them with the hormone levels observed in these patients: the levels of LH and FSH in this group are systematically greater than 50 m.I.U./ml. (Table V)

131

It would seem from the above data that we may conclude that the presence of laparoscopic findings of ovarian hypoplasia is evidence of primary or secondary ovarian insufficiency in 100% of cases.

When the same type of analysis is carried out for PCOS, marked by an LH/FSH ratio of   2, we can see how laparoscopic finding of PCO is confirmed by the hormone assay in 71% of cases, while the latter is supported by a corresponding ovarian appearance in 77% of cases (Table VI).

TABLE V Laparoscopic reports of ovarian hypoplasia compared with gonadotropin levels below or above normal.

	LH and FSH $<$ 10 mIU/ml	LH and FSH $>$ 50 mIU/ml
Ovarian hypoplasia	42%	58%
Streak-like ovary	– –	100%

TABLE VI Relationship between laparoscopic reports of polycystic and smooth ovaries and LH/FSH ratios of $\geq$ 2.

100%	P.C.O.	LH/FSH $\geq$ 2
LH/FSH $\geq$ 2	77% (vs 23%: other laparoscopic findings)	– –
P.C.O.	– –	71% (vs 29%: LH/FSH $\geq$ 2)

It should further be noted that as the laparoscopic finding progresses from the smooth ovary to the polycystic ovary and, finally, to the Stein-Leventhal-type ovary,

correlation with the hormone situation increases, passing
from 56% for the former to 85% for the latter (Table VII).

TABLE VII Correlation between laparoscopic reports similar to
PCOS and laboratory data.

Laparoscopic reports	LH/FSH $\geq$ 2	LH/FSH $\geq$ 2
Smooth ovary	56%	44%
Polycystic ovary	75%	25%
Stein ovary	85%	15%

In this type of pathology there would therefore seem
to be considerable similarity between laparoscopic findings
and hormone assay data, suggesting a good degree of correlation between these two investigations.

LAPAROSCOPY IN THE DEFINITION OF TREATMENT

Laparoscopy is also of considerable assistance in
the chjoice of treatment, enabling the most suitable drug
for inducing ovulation to be selected on the basis of the
ovarian report. It is well known that clomiphene gives
excellent results in women affected with polycystic ovaries
while it is of little or no therapeutic effect in situations
of marked ovarian hypoplasia. In these cases, on the
other hand, effective treatment can be obtained with preparations of gonadotorpins or through the administration of GnRH.

In gonadotropin-based treatment, moreover,
endoscopic evaluation of the ovarian volume permits more
accurate selection of the dosage, also indicating
parameters for conduct of the ovarian stimulation.
In fact, if we examine cases of ovarian hypoplasia in
which normal ovulation has been achieved in our experience
with gonadotropins, it can be seen that there is a direct
and linear correlation between the laparoscopic report and
the initial and total numbers of phials administered,
the duration of treatment and the plasma estradiol levels
at which therapy was suspended (Table VIII).

TABLE VIII Relationship between laparoscopic report and parameters for ovarian stimulation.

Laparoscopic report	Initial no. $\bar{X}$	Total no. $\bar{X}$	Days of Treatment	E2
Mild hypoplasia	2.5	25	7.8	555±240
Moderate hypoplasia	3.8	44	9.3	630±258
Severe hypoplasia	5.8	77	11.0	945±370

It is thus essentially possible, without forgetting the fundamental principle that this type of therapy must be personalised, to draw from these data useful indications for a correct conduct of ovulation induction with gonadotropins.

CONCLUSIONS

In conclusion, it may be said that laparoscopy maintains a role of primary importance in the search for diagnosis and treatment of infertile patients, often affected with dysendocrinia. In this context the gynaecologist and endoscopist cannot withdraw from study of the ovary in all its many morphological aspects in order to acquire physiopathological information. The ovary, with its ability to relate the present while recalling the past, can be a valuable informer of a patient's endocrine situation.

REFERENCES

1. Steptoe PC. Leparoscopy in gynaecology. Livingstone Ed. Edinburgh, 1967
2. Semm K. Atlas de coelioscopie ed d'hystéroscopie. Ed. Masson, Paris, 1977
3. Loffer FD, Pent D. Indications—contraindications and complications of laparoscopy. Obstet Gynecol Surv, 30:407, 1975
4. Hamou J. Hystéroscopie et microcolpohystéroscopie. Atlas et traité. Ed. COFESE, Palermo, 1984
5. Hamou J. Taylio PJ. Panoramic, contact and microcolpohysteroscopy in gynecological practice. Curr Prob Obstet Gynecol Year Book, 1982

134

Morphological features of aspirated human oocytes in an IVF program

C. ARAGONA, G. MICARA

2nd Dept. of Obstetrics and Gynecology (Chairman Prof. L. Carenza), IVF Program, University "La Sapienza", Rome, Italy

INTRODUCTION

In Vitro Fertilization (IVF) Programs provide a unique opportunity to correlate morphology and function of human oocytes.Most teams throughout the world induce multiple follicular development by the administration of Clomiphen citrate and/or gonadotropins combined in a number of protocols whose efficacy mostly rely on the relative experience of the investigators.
Important changes occur during follicular maturation that are brought about by gonadotropins and possibly by estradiol as the follicles mature from the primordial stage through the graafian stage. Ultrasounds clearly show the increase of the diameter and the volume of the large follicles still growing when hMG administration is discontinued so that it is possible to represent the increase of total follicular volume.When we look at the peripheral hormonal situation in Clomiphen citrate–hMG–hCG stimulated cycles of patients who concieved in our IVF Program,fixing the reference point provided by the hCG administration or the LH surge,it is possible to observe in the periovulatory period:

1) an increase in LH activity peaking at 12-24 hours following the hCG administration
2) an increase of 17B–estradiol followed by a slow drop that seems to be more linked to the decrease in the aromatisable androgens consequent to the modifications in the biogenesis of the theca rather than to modifications of the capacity for aromatization of the granulosa
3) an increase in progesterone whose quantity in the follicle multiplies by 10 before ovulation.

135

Such peripheral hormonal pattern allows us to distinguish four functional areas that appear to be of some practical importance to pin point the time of follicular aspiration:

STAGE 1= IMMATURITY
STAGE 2= PREMATURITY
STAGE 3= MATURITY
STAGE 4= POSTMATURITY

The aim of the present study is to evaluate more precisely the mor_phological features of human oocytes aspirated at stage 3(maturity).

MATERIALS AND METHODS

The morphological features of human oocytes evaluated by an inver_ted brightfield microscope at magnification of 25 to 100x are initially reported at the time of aspiration in a stimulated cycle. The first classification is definitely essential to determine the stage of maturation of each oocyte,in order to assess the ideal mo_ment of insemination(1,2).The criteria of classification are based on the granulosa cells appearance combined with the characteristics of the oocyte-corona radiata-cumulus complex(3).The features cha_racterizing the different stages of maturation are:

GERMINAL VESICLE:present in the immature oocyte
FIRST POLAR BODY:present and occasionally visible in the mature
                  oocyte
CORONA RADIATA:1)compact or absent in the immature oocyte
               2)expanded in the mature oocyte
CUMULUS:1)absent or small in the immature oocyte
        2)present and expanded in the mature oocyte
GRANULOSA CELLS:1)non-luteinized in association with the immature
                  oocyte
                2)luteinized in association with the mature oocyte.

A third kind of oocyte has been observed and classified as post-ma_ture when an oocyte apparently mature showed a dark ooplasm and a thickness of the zona pellucida.When a fractured zona pellucida is observed the oocyte can be classified as degenerated;in these cases a mechanical damage,related to hand aspiration,appears to be involved.

Immature oocytes are in various stages of the first meiotic divi_sion and pre-incubated for 24-30 hours for the completion of meiotic maturation prior to insemination in vitro(4).
Meiotically mature oocytes are inseminated after 4-6 hours of pre-incubation.Post-mature oocytes,depending on the progress of degene_ration,are generally non viable.

136

<u>RESULTS</u>

The experience acquired in the last two yearscombining the periphe_
ral hormonal pattern with the morphological aspect of the oocyte at
aspiration,drove us to the conclusion that a more careful attention
to the oocyte meiotic stage must be applied.The resumption of the
first meiotic division and the final maturation of the oocyte with
the extrusion of the first polar body (metaphase II) must be obser_
ved in order to determine a more correct time for insemination.
Four different stages of maturation can be identified:

STAGE 1:immature oocyte with a germinal vesicle still present after
        24-30 hours of pre-incubation,failing maturation (Fig.1)
STAGE 2:premature oocyte with a germinal vesicle at aspiration ca_
        pable of resuming meiosis up to metaphase II.The first po_
        lar body extrusion will allow us to inseminate (Fig.2)
STAGE 3:mature oocyte.Four different stages of maturation can be
        observed depending on the expansion of the corona radiata:
        A-B.slightly compact without   clear evidence of polar body
        (Fig.3)
        C.well expanded with a polar body and a regular and light
        ooplasm (Fig.4)
        D.well expanded with a polar body but a dark ooplasm and
        thickness of the zona pellucida (Fig.5)
STAGE 4:post-mature oocyte as previously described (Fig.6).

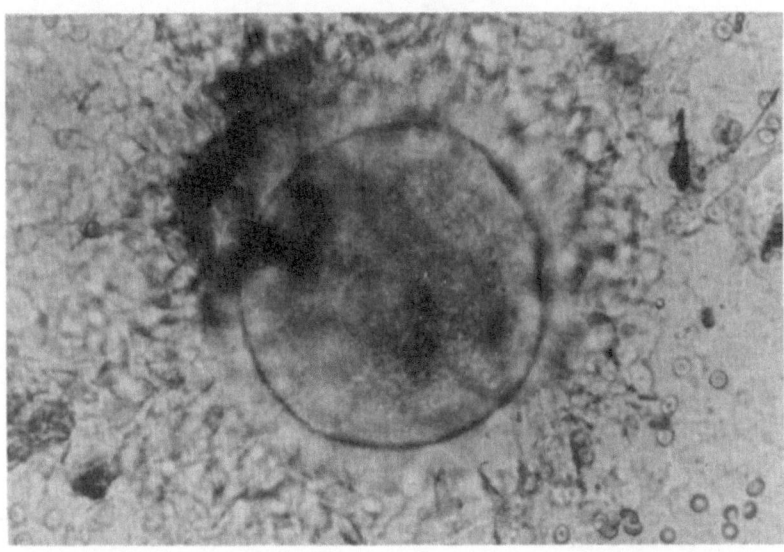

Fig.1.Oocyte at stage 1.Germinal vesicle did not breakdown after
      24 hours of pre-incubation in vitro.

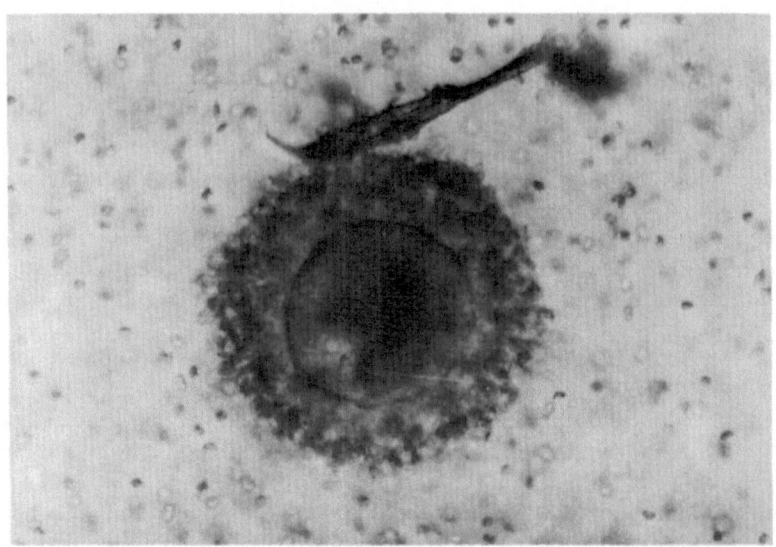

Fig.2.Oocyte at stage 2.Compact corona radiata and germinal vesicle
that will breakdown in 24-30 hours.First polar body extrusion
will follow.

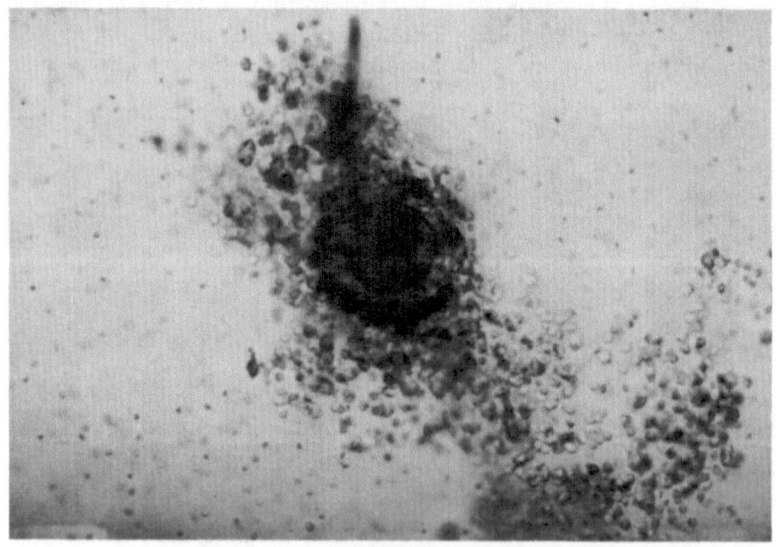

Fig.3.Oocyte at stage 3A-3B.Expanded cumulus,slightly compact coro_
na.The oocyte will be inseminated as soon as the first polar
body will be extruded.

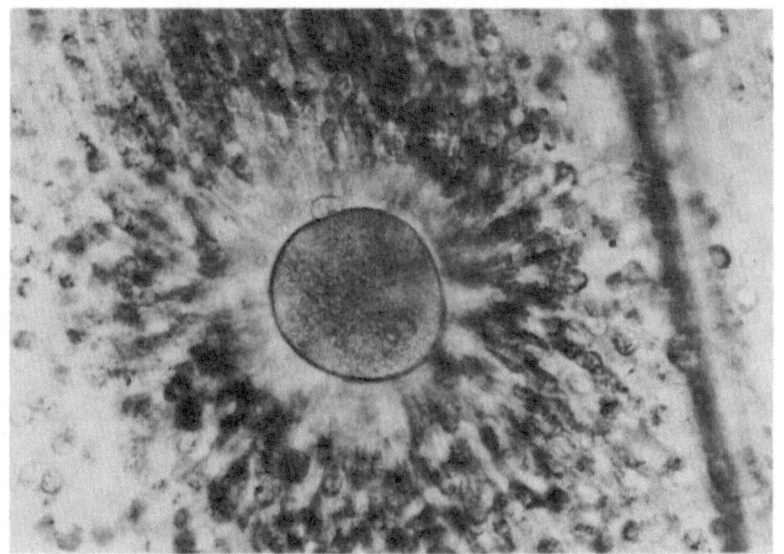

Fig.4.Oocyte at stage 3C.Expanded cumulus, expanded corona,visible
first polar body and light ooplasm.

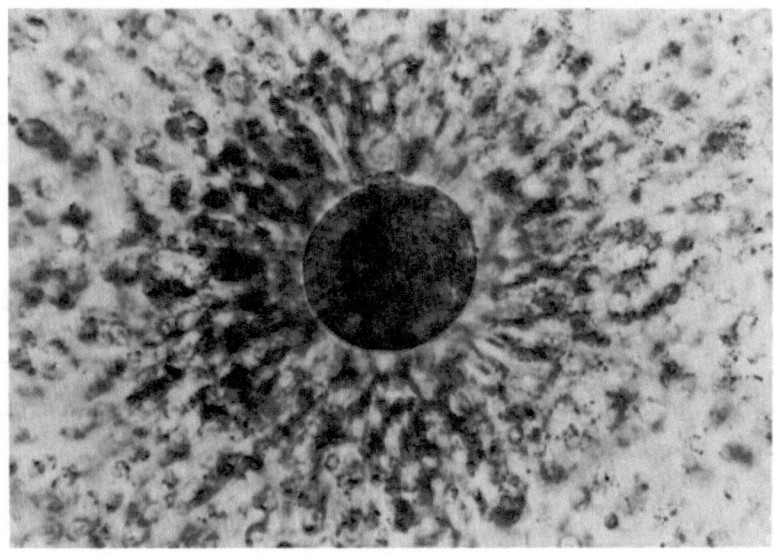

Fig.5.Oocyte at stage 3D.Dark ooplasm and | thickness of the zona
pellucida

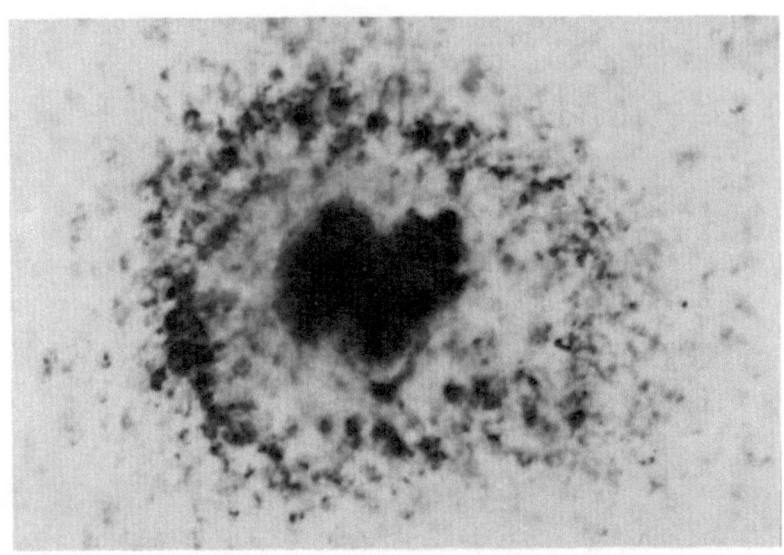

Fig.6.Degenerated oocyte with fractured zona pellucida and dis_
     persion of the ooplasm.

Most of the oocytes are generally in stage 3 and a correct identifi
cation of the maturation sub-stage allows us to pre-incubate these
oocytes up to the moment of a visible polar body (1 to 12 hours).
A delay in the insemination can be consistent with polyspermy or no
fertilization (5) as well as an early insemination wont allow ferti
lization.A correct time of insemination in a suitable culture me_
dium combined with the quality of the oocyte at aspiration seems to
influence the fertilization rate (Tab.I) and embryo quality.

Tab.I.Fertilization rate of human oocytes in different morphologi-
     cal maturation stages

Morphological stage at aspiration		N°oocytes	% of fertilization
1		22	0
2		70	25
3	3A	13	30
	3B	60	40
	3C	280	80
	3D	35	20
4		20	5

The criteria for embryo classification are based on the time of cleavage,the shape of blastomers and the appearance of cytoplasm before transfer.

DISCUSSION

Is the quality of the oocyte related to the quality of the hormonal environment?The retrieval of oocytes at different stages of matura_ tion in the same patient is the best answer at that question.On the other hand if the hormonal levels in the blood do not reflect the hormonal concentration in the follicular fluids (6) it is equally true that such investigations do not allow to predict the fertili_ zation and cleavage patterns of human oocytes.
In the discussion of what parameters in the follicle are required to insure that an egg will be mature and fertilizable and lead to a normal pregnancy,two aspects can not be underestimated:
1)the role of some factors present in the follicular fluid which are probably involved in the follicle-oocyte maturation such as the inhibin,the OMI,the luteinization inhibit ing factors,the FSH binding inhibin, the gonadocrinins(7);
2)the role played by the cumulus-corona radiata complex and the zo_ na pellucida in the exchanges ex isting between the oocyte and the follicular fluid.
Data obtained by scanning and transmission electron microscopy and reported elsewhere in this publication (Familiari et al.) compared with light microscopical observations and peripheral hormonal pat_ terns can provide further useful morphological details for correc_ tly evaluating human oocytes at different stages of maturation.

CONCLUSIONS

In conclusion our study clearly indicates that an improvement in the morphological classification of human oocytes can be performed at aspiration.Our preliminary data seem to show that fertilization rates can largely depend upon the stage of maturation,being stage 3C the most satisfactory (up to 80%).Certainly further studies are needed to confirm, on a larger number of oocytes,our observations and to demonstrate that modifications of the timing of incubation in relation to the morphological features can be of practical uti_ lity in improving fertilization rates and embryo quality.

REFERENCES

1) Sandow BA:Characteristics of human oocytes aspirated for in Vitro Fertilization.In:Infertility,ed. Acosta ÁA, New York and Basel, Dekker M. Inc.,1983:6(1-4),143.

2) Plachot M,Mandelbaum J,Cohen JJ,Debache C,Pigeau F and Junca AM: Sequential use of Clomiphene citrate,human menopausal gonadotropin, and human chorionic gonadotropin in human in Vitro Fertilization.I. Follicular growth and oocyte suitability.Fertil Steril 1985:43,255.

3) Eppig JJ:The relationship between cumulus cell-oocyte coupling, oocyte meiotic maturation and cumulus expansion.Dev Biol 1982:89, 268.

4) Veeck LL,Wortham JWE Jr,Witmyer J,Sandow BA,Acosta AA,Garcia JE, Jones GS,Jones HW Jr:Maturation and fertilization of morphologi_ cally immature human oocytes in a Program of in Vitro Fertiliza_ tion.Fertil Steril 1983:39,594.

5) Trounson AO,Mohr LR,Wood C,Leeton JF:Effect of delayed insemina_ tion on in Vitro Fertilization,culture and transfer of human embryos.J Reprod Fertil 1982:64,282.

6) Flower RE,Edwards RC,Walters DE,Chan STH,Steptoe PC:Steroidogene_ sis in preovulatory follicle of patients given human menopausal and chorionic gonadotropins as judged by the radioimmunoassay of steroids in follicular fluid.J Endocr 1978:77,161.

7) Channing CP,and Tsafriri A:Mechanism of action of luteinizing hormone and follicle-stimulating hormone on the ovary in vitro. Metabolism 1977:26,413.

142

Proteins in the structure and function of human spermatozoa

B. BACCETTI

Department of Evolutionary Biology, University of Siena, Siena, Italy

The spermatozoon is one of the cell models where results obtained by chemical analysis have better corresponded with morphological descriptions offered by electron microscopy, so that for the most important and peculiar sperm organelles,form and composition are closely related. This result is very important for human pathology, where absence, due to various reasons, of particular proteins is concomitant with absence or malformation of organelles. The classical approach to human sperm structure and pathology has been the electron microscopy examination (Pedersen, 1974; Fawcett, 1975; Baccetti, 1978), that is still now the most useful technique, but chemical identifications must be performed in parallel on the same material, because an organelle is usually made up by different polypeptides. On the contrary, protein analysis in toto on spermatozoa gives poor results (Mujica et al., 1978) because the same protein is frequently present on different organelles, and takes part in different functions. So, at this moment, sperm investigation is a more and more difficult problem, needing the contemporaneous utilization of submicroscopical and biochemical techniques. Here a brief comment on the "status of art" of this complex and multidisciplinary matter.

1. The acrosomal enzymes

The acrosin is the most important component of the acrosome (Fig. 1): it is a typical enzymatic protein 30.000 daltons m.w., responsible for the penetration of the zona pellucida. Apparently it is very conservative, being present at any level of the animal kingdom (Baccetti, 1979), but in humans, at the time of ejaculation it is present, for the 90%, in the form of precursor (proacrosin) which is inactive for the presence of an inhibitor (acrostatin). Human acrosin can be localized by immunofluorescent antibodies raised against the acrosin of other mammals (Garner and Easton, 1977): evidently the presence of the precursor renders this technique not so exhaustive, and more specific detect ion of protein, precursor and inhibitor must be developed, provided Goodpasture et al. (1980) succeeded in extracting the three components from human sperm.

Hyaluronidase is another important acrosomal protein 60,000 daltons m.w., consisting of 4 polypeptides (McRorie and Williams, 1974), acting on the cumulus complex of the oocyte. De Vries et al. (1985) are able to localize it in porcine spermatozoa by immunocytochemical staining. A third enzyme is the collagenase, 110 daltons m.w., found in the acrosome of several mammals, including humans, by Koren and Milkovic (1973), never localized cytochemically. Evidently research must be performed in order to complete the enzyme topography in human acrosome by appropriate antibodies used as molecular probes. This technique will be useful in pathology, given the frequent presence of acrosomal malformations, and even complete acrosomal absence, in human ejaculates.

2. The problem of actin and myosin.

Myosin has been detected by immunofluorescent staining by Clarke and Yanagimachi (1978) and by Campanella et al. (1979) in human acrosomal region, while Virtanen et al. (1984) have been able only to localize it in the neck region. Actin also has been localized in the human acrosomal region by Clarke and Yanagimachi (1978) and by Campanella et al. (1979), while Talbot and Kleve (1978) found it only in the tail region of the hamster sperm. Clarke et al. (1982) suggested that the actin is present in monomeric form in the head, and as F-actin in the tail. On the contrary, Flaherty et al. (1983) bind phallacidin to the acrosome of the Plains Mouse, concluding for the presence of F-actin in this region. Baccetti et al. (1984) found actin in the spiral sheath and in the pericentriolar region of bull sperm, result obtained also by Virtanen et al. (1984a) who bind antiactin antiserum to human acrosome and tail, but fail to stain with phallacidin, and suspect to be in presence of G-actin in any case. Also Welch and O'Rand (1985) find non filamentous actin in rabbit post acrosomal region, while Saxena et al. (1986) observe that boar spermatozoa develop actin filaments in the postacrosomal region and in the flagellum only after capacitation. No doubt, therefore, that human spermatozoa really contain actin (Ochs and Wolf, 1985), which is present mainly in the post acrosomal, subacrosomal (Fig. 1), centriolar and spiral sheath (Figs. 2,3) regions. As to the problem of G- and F-actin the situation is still confused, even if the suggestion of Saxena et al. (1986) appears very interesting. Quite recently Camatini et al. (1986a) suggest that Y-actin could be present only around the acrosome, β-actin in other regions. A new problem, thus arises. This acto-myosin system is probably part of the filamentous

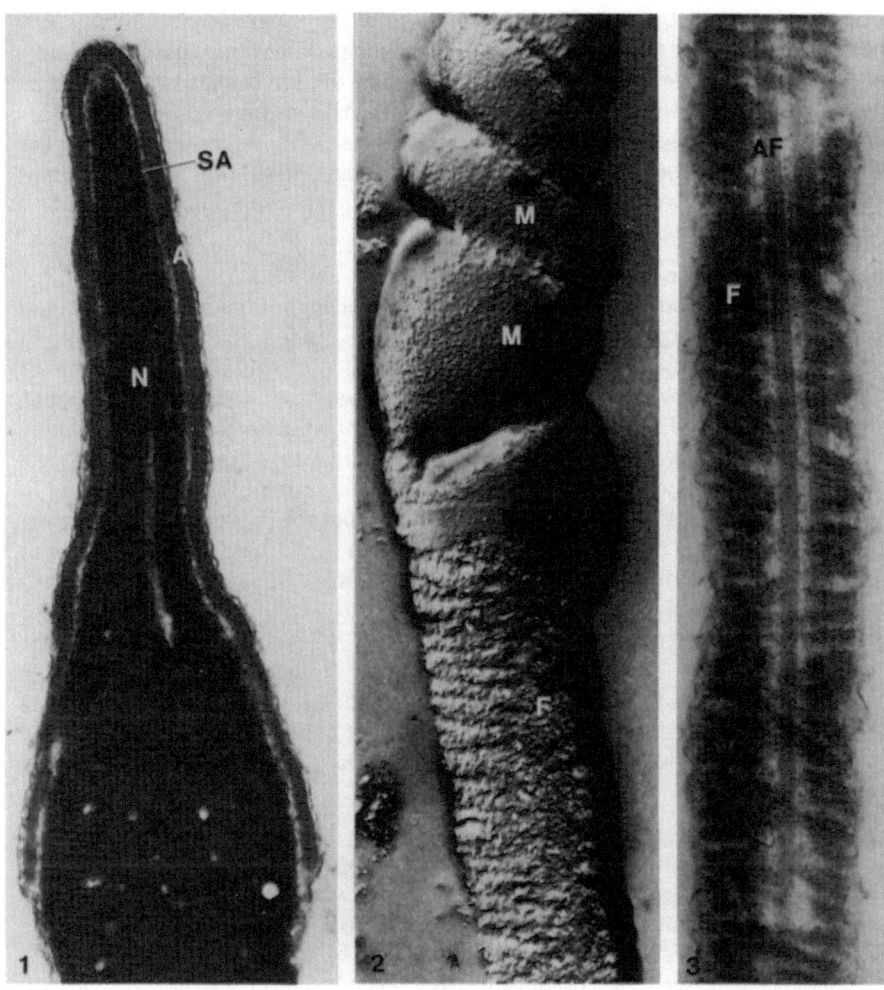

Fig. 1 - Longitudinal section of human sperm head showing the acrosome
(A) and the compact nucleus (N). Between them, in the
subacrosomal space (SA) a thin layer of filamentous material
containing actin filaments is evident. Philips 301 T.E.M.,
45,000 X.

Fig. 2 - Frozen-etched preparation of the tail region in a human
spermatozoon, showing the spiral arrangement of the
mitochondrial helix (M) and of the fibrous sheath (F)
containing actin. Philips 301 T.E.M., 45,000 X.

Fig. 3 - Longitudinal section of the principal piece in a human
spermatozoon showing the spiral organization of the fibrous
sheath (F) surrounding the accessory fibers (AF). Philips 301
T.E.M., 75,000 X.

network observed by Escalier (1984) in human spermatozoa. It seems therefore, to be implicated in the acrosomal reaction and to play a role in the movement, or in the maintaining of tail shape, as part of the cytoskeleton in the centriolar and in the principal piece region of the sperm. Moreover, an intriguing datum is the presence of actin in the inner axonemal arm (Piperno and Luck, 1979) of Chlamydomonas flagella. If confirmed in sperm, this localization could suggest a possible role of actin in axonemal motility.

3. The intermediate sized filaments.

The 10 nm filaments are the last proteinaceous structures localized in human spermatozoa. All research has been carried out with specific antibodies. Firstly Baccetti et al. (1984) and Virtanen et al. (1984a) detected vimentin in the mammalian (also human) post acrosomal sheath (Fig. 1); the first AA. were also able to localize, in the same sperm, keratin filaments in the satellite fibrils of the tail. More recently Ochs et al. (1986) found vimentin and keratin in the human acrosomal region, changing distribution after the acrosomal reaction. It is, thus ,possible that some of the cross-filaments observed by Escalier (1984) by electron microscopy belong to the intermediate sized filaments category. No other information is available, but certainly the distribution of these filaments in normal and pathological sperm needs further investigation . Another intermediate sized filamentous protein, tektin, localized by Linck et al. (1982) in the flagellar microtubules of sea urchin, is possibly present also in mammalian spermatozoa. This aspect needs further clarification.

4. The tubulin-dynein system.

Tubulin (Figs. 4,6,7) is not a uniform widely distributed protein dimer as suspected at earlier times. At first, in algae and protozoa, subsequently in animal sperm cell multiple form of tubulin and different domains, along a single microtubule, have been demonstrated. This heterogeneity is important for the stability of flagellar microtubule (Kobayashi, 1982), as well as for the specific domain on the tubule A involved in dynein binding (Haimo and Rosenbaum, 1981). A differential localization of α and β tubulin in human spermatozoa has been demonstrated by Virtanen et al. (1984b); more recently Gallo et al. (1986) find, in addition to the obvious presence of tubulin in the centriole and axoneme regions, a peculiar distribution of a tubulin epitope, different in the various spermatozoa of the same donor and possibly related to abnormalities of the peri-axonemal structures in pathological cases. Analogous is the problem of dynein. Several dynein bands (at least 4) have been detected in human axoneme (Baccetti et al., 1981), localized in the inner and outer arms (Figs. 4,6,7), near the head of the radial spokes and on the peripheral Y links. Pathological spermatozoa, deficient of motility and of some of these

146

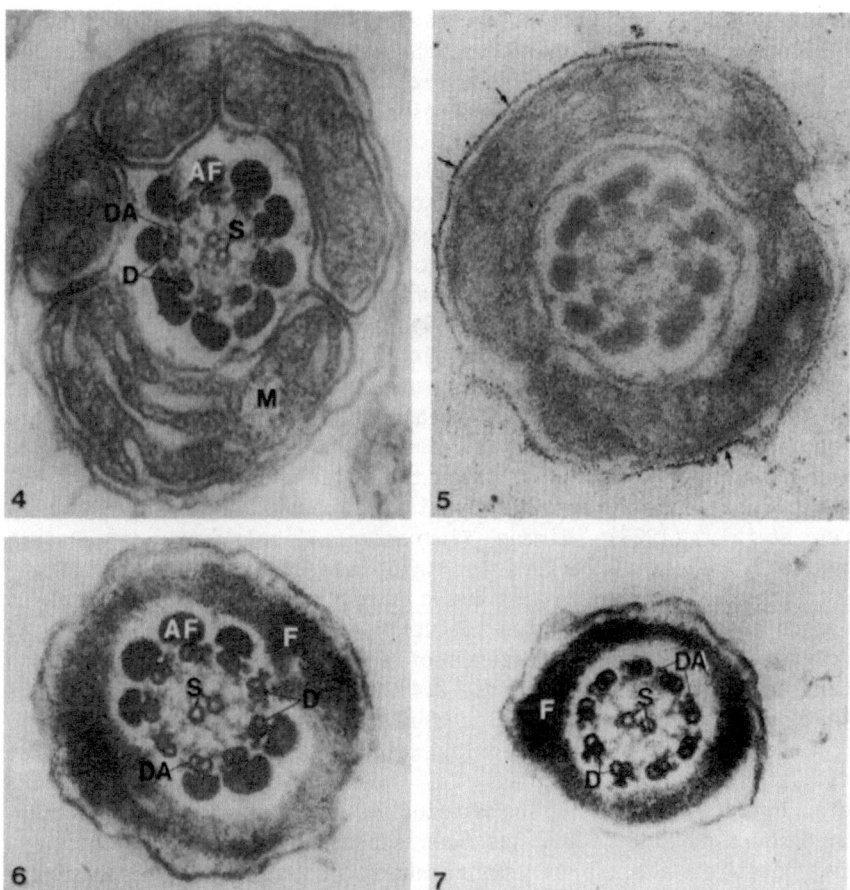

Fig. 4 - Cross section of the midpiece in a human spermatozoon showing
         mitochondria (M), accessory fibers (AF) and, in the axoneme,
         tubulin doublets (D) and singlets (S), the first endowed with
         dynein arms (DA). Philips 301 T.E.M., 100,000 X.

Fig. 5 - The same section, after Thiéry method staining glycoproteins
         on the plasma membrane (arrows). Philips 301 T.E.M., 100,000
         X.

Fig. 6 - Cross section of the principal piece in a human spermatozoon,
         showing the actin containing fibrous sheath (F), the
         diminished in number accessory fibers (AF) and the tubulin
         singlets (S) and doublets (D) with dynein arms (DA). Philips
         301 T.E.M., 100,000 X.

Fig. 7 - The same section in a more posterior region devoid of
         accessory fibers. Same indications as in Fig. 6. Philips 301
         T.E.M., 100,000 X.

structures, also specifically lack one or more of these polypeptides.
5. The parergins complex and other S-S stabilized proteins.

Accessory fibers of human spermatozoa (Figs. 3,4,6) are made up of a class of polypeptides, the parergins, belonging to two different groups (Baccetti et al., 1976): one of relatively high m.w. (80,000-50,000 daltons) polypeptides with a high content of leucine and glutamic acid, and a low content of cysteine and proline, forming helices of 2 filaments 2 nm thick; the second of low m.w. polypeptides (31,000-28,000 daltons), low in leucine and glutamic acid, high in cysteine and proline, forming the amorphous part of the fiber, binding zinc by reaction with sulphydryl groups and cross linked to the first chains by disulfide bridges. Subsequently Vera et al. (1984) add that the major chains are phosphorylated at serine residues, and suggest parergins to be a unique family of phospho-proteins. No anti-parergin antibodies are available as yet; their preparation will enable to investigate on large number of cases the relationship between abnormal length of accessory fibers and dyskinesia demonstrated in humans by Serres et al. (1986). Bimane fluorescent labels are at present sufficiently selective markers for thiol and disulfide groups (Huang et al., 1984). With this technique a specific action of gossypol on S-S bonds has been demonstrated (Baccetti et al., 1986). This staining could also account for the presence of another S-S stabilized protein present around mitochondria (Figs. 2,4) (Pallini et al., 1979).
6. Calmodulin, spectrin and the problem of calspectin.

Calmodulin is a low molecular weight, heat stable protein that binds Ca^{2+} and regulates many Ca^{2+} dependent enzymes, including dynein. Detected before in spermatozoa of sea urchin (Jones et al., 1978; Garbers et al;, 1980), has been subsequently found in mammalian sperm. Jones et al. (1980) and Feinberg et al. (1981) demonstrate calmodulin to be present in the head and tail region of different mammalian spermatozoa. More precisely, Gordon et al. (1983), using a ferritin-labeled antibody, localize by electron microscopy, calmodulin close to the tubulin-dynein system of guinea-pig axoneme, while Camatini et al. (1986b) find calmodulin associated with plasma membrane and outer acrosomal membrane in boar sperm heads, and the structure of outer acrosomal membrane to be particularly sensitive to anticalmodulin drugs (Berruti et al., 1985).

Calmodulin is bound to a particular protein, the α-spectrin, localized first in erythrocytes, after in many other cells, except spermatozoa (Repasky et al.,1982). So the problem of the binding protein for sperm calmodulin remains open. Subsequently Virtanen et al. (1984a) detect α-spectrin in the acrosomal and tail regions containing calmodulin in human spermatozoa. Spectrin is known also to bind G-actin, that is widely diffused, as we have seen, in the same region. Tsukita et al. (1983) find that the rod-like dimers of the

148

brain spectrin (calspectin) bind calmodulin in a region of the rod, actin in another. Calspectin of spermatozoa needs to be detected, being a signal of presence and status both of actin and calmodulin.

7. Nucleoproteins.

The basic proteins that in human spermatid substitute the somatic histones present in spermatogonia are protamines and histones rich in arginine (about 50%) and cystine (about 10%). Kolk and Samuel (1975) have isolated from human sperm nuclei (Fig. 1) two proteins, protamine 1 and 2, of m.w. 6280 and 6840 daltons respectively, but more recently Gusse et al. (1984) were able to distinguish 7 fractions of protamines and 2 intermediate fractions between histones and protamines, and Dadoune and Alfonsi (1986) suggest in most of human ejaculates the presence of spermatozoa with persisting basic proteins related to a disturbance of chromatin condensation. A particular human abnormality, the round headed spermatozoa, involves a poor formation of S-S bonds (Baccetti et al., 1977), and also Gossypol interferes in the same process (Baccetti et al., 1986). Nucleoproteins and male sterility need to be comparatively examined.

8. Membrane proteins.

The use of lectins and other probes has demonstrated the presence on the sperm membrane of glycoproteins (Fig. 5) variously distributed in the head and in the tail regions. This glycoprotein layer is important because it contains surface antigens with specific localization, as well as sites interacting with ATP and cAMP, undergoing conformational changes that facilitate the transport of substrates, influencing capacitation (Mercado et al., 1974) and initiation of movement (Morisawa, 1986). It contains also sialic acid, which influences the difference in negative charge between X- and Y-bearing human sperm (Kaneko et al., 1984). Virtanen et al. (1984a) find on human sperm surface different localizations for the various lectin conjugates, sometimes related to the distribution of the peripheral cytoskeleton (vimentin and actin). This is in good agreement with the results obtained with the electron microscope by Escalier (1984). Submembranary cytoskeleton and sperm surface polypeptides are, thus, demonstrated to be linked, and this view will provide a good approach in order to understand the movement of membrane proteins in the various moments of sperm maturation and activity, as well as their influence in pathological conditions or after drug administration.

This complex of data demonstrates that the recent effort of cell biologists to extend to human spermatozoon the many observations firstly obtained on different animal models is giving a convincing reconstruction of the molecular structure of this cell, and opens interesting approaches to fertility-infertility control.

149

References

Baccetti B. Lo spermatozoo umano. In: I° Congresso Naz. Società Ital. Andrologia, Relazioni. ed. Pacini. 1978: 123-167.

Baccetti B. The evolution of the acrosomal complex. In: The spermatozoon, eds. Fawcett DW, Bedford JM Baltimore-Munich, Urban & Schwarzenberg 1979: 305-329.

Baccetti B, Pallini V, Burrini AG. The accessory fibers of the sperm tail. II. Their role in binding zinc in Mammals and Cephalopods. J Ultrastruct Res 1976: 54, 261-275.

Baccetti B, Renieri T, Rosati F, Selmi MG, Casanova S. Further observations on the morphogenesis of the round headed human spermatozoa. Andrologia 1977: 9, 255-264.

Baccetti B, Burrini AG, Pallini V, Renieri T. Human dynein and sperm pathology. J Cell Biol 1981: 88, 102-107.

Baccetti B, Bigliardi E, Burrini AG, Gabbiani G, Jockusch BM, Leoncini P. Microfilaments and intermediate sized filaments in sperm tail. J Submicrosc Cytol 1984: 79-84.

Baccetti B, Bigliardi E, Burrini AG, Renieri T, Selmi MG. The action of gossypol on rat germinal cells. Gam Res. 1986: 13, 1-17.

Berruti G, Anelli G, Camatini M. The effects of anticalmodulin drugs on the ultrastructure of boar spermatozoa. Eur J Cell Biol 1985: 39, 147-152.

Camatini M, Casale A, Cifarelli M., Corno A, Anelli G. Immunoriconoscimento di isoforme di actina non-muscolare in spermatozoi di verro. Atti V° Congresso ABCD, 1986a, 108.

Camatini M, Anelli G, Casale A. Immunocytochemical localization of calmodulin in intact and acrosome-reacted boar sperm. Eur J Cell Biol 1986b: 41, 89-96.

Campanella C, Gabbiani G, Baccetti B, Burrini AG, Pallini V. Actin and myosin in the vertebrate acrosomal region. J Submicrosc Cytol 1979: 11, 53-71.

Clarke GN, Yanagimachi R. Actin in mammalian sperm heads. J exp Zool 1978: 205, 125-131.

Clarke GN, Clarke FM, Wilson S. Actin in human spermatozoa. Biology of Reproduction 1982: 26, 319-327.

Dadoune JP, Alfonsi MF. Ultrastructural and cytochemical changes of the head components of human spermatids and spermatozoa. Gam Res 1986: 14, 33-46.

De Vries JWA, Willemsen R, Geuze HJ. Immunocytochemical localization of acrosin and hyaluronidase in epididymal and ejaculated porcine spermatozoa. Eur J Cell Biol 1985: 37, 81-88.

Escalier D. The cytoplasmic matrix of the human spermatozoon: cross-filaments link the various cell components. Biol Cell 1984: 51, 347-364.

150

Fawcett DW. The mammalian spermatozoon. Develop Biol 1975: 44, 394-436.

Feinberg J, Weinmann J, Weinmann S, Walsh MP, Harricane J, Gabrion J, Demaille JG. Immunocytochemical and biochemical evidence for the presence of calmodulin in bull sperm flagellum. Biochim Biophys Acta 1981: 673, 303-311.

Flaherty SP, Breed WG, Sarafis V. Localization of actin in the sperm head of the plains mouse Pseudomys australis. J Exp Zool 1983: 225, 497-500.

Gallo JM, Escalier D, Schrevel J, David G. Differential distribution of tubulin epitopes in human spermatozoa. Eur J Cell Biol 1986: 40, 111-116.

Garbers DL, Hansbrough JR, Radany EW, Hyne RV, Kopf GS. Purification and characterization of calmodulin from sea urchin spermatozoa. J Reprod Fert 1980: 59, 377-381.

Garner DL, Easton MP. Immunofluorescent localization of acrosin in mammalian spermatozoa. J exp Zool 1977: 200, 157-162.

Goodpasture JC, Polakoski KL, Zaneveld LJD. Acrosin, proacrosin and acrosin inhibitor of human spermatozoa: extraction, quantitation and stability. J Andrology 1980: 1, 16-27.

Gordon M, Morris EG, Young RJ. The localization of Ca^{2+}-ATPase and Ca^{2+} binding proteins in the flagellum of Guinea Pig sperm. Gam res 1983: 8, 49-55.

Gusse M, Sautière P, Roux Ch, Dadoune JP, Chevaillier Ph. Extraction, purification et caractérisation des protéines nucleaires du spermatozoide humain. Biol Cell 1984: 52, 93a.

Haimo LT, Rosenbaum JL. Dynein binding to microtubules containing microtubule-associated proteins. Cell Motility 1981: 1, 499-516.

Huang TTF, Kosower NS, Yanagimachi R. localization of thiol and disulfide groups in Guinea Pig spermatozoa during maturation and capacitation using bimane fluorescent labels. Biol Reprod 1984: 31, 797-809.

Jones HP, Bradford MM, McRorie RA, Cormier MJ. High levels of a calcium-dependent modulator protein in spermatozoa and its similarity to brain modulator protein. Biochem Biophys Res Commun 1978: 82, 1264-1272.

Jones HP, Lenz RW, Palevitz BA, Cormier MJ. Calmodulin localization in mammalian spermatozoa. Proc Natl Acad Sci USA 1980: 77, 2772-2776.

Kaneko S, Oshio S., Kobayashi T, Iizuka R, Mohri H. Human X- and Y-bearing sperm differ in cell surface sialic acid content. Biochem Biophys Res Comm 1984: 124, 950-955.

Kobayashi Y. Stable microtubules in starfish sperm flagellum: their structures and heterogeneity of tubulin. J Biochem 1982: 92, 1305-1318.

Kolk AHJ, Samuel T. Isolation, chemical and immunological charcterization of two strongly basic nuclear proteins from human spermatozoa. Biochem. Biophys Acta 1975: 393, 307-319.

Koren E, Milkovic S. Collagenase like peptidase in human, rat and bull spermatozoa. J Reprod Fert 1973: 32, 319-356.

Link RW, Albertini DF, Kenney DM, Langevin GL. Tektin filaments: chemically unique filaments of sperm flagellar microtubules. Cell Motil 1982: suppl 1, 127-132.

McRorie RA, Williams WL. Biochemistry of mammalian fertilization. Ann Rev Biochem 1974: 43, 777-803.

Mercado E, Hicks JJ, Drago C, Rosado A. A study of the interaction of human spermatozoa membrane with ATP and cyclic-AMP. Biochem Biophys Res Communic 1974: 56, 185-192.

Morisawa M. The process of the initiation of sperm motility. In: Developm. Growth & Different. New horizons in spermatozoa research 1986: 28, suppl., 13-18.

Mujica A, Alonso R, Hernandez-Montes H. Electrophoretic patterns of total, nuclear, and flagellar proteins from ejaculated human spermatozoa. Int J Fertil 1978: 23(2), 112-117.

Ochs D, Wolf DP. Actin in ejaculated human sperm cells. Biol reprod. 1985: 33, 1223-1226.

Ochs D, Wolf DP, Ochs RL. Intermediate filament proteins in human sperm heads. Exptl Cell Res 1986: 167, 495-504.

Pallini V, Baccetti B, Burrini AG. A peculiar cysteine-rich polypeptide related to some unusual properties of mammalian sperm mitochondria. In: The Spermatozoon, eds. Fawcett DW, Bedford JM Baltimore-Munich, Urban & Schwarzenberg 1979: 141-151.

Pedersen H. The human spermatozoon; thesis Aarhus University. Danish Med Bull 1974: 21, 1-48.

Piperno G, Luck DJL. An actin-like protein is a component of axonemes from Chlamydomonas flagella. J Biol Chem 1979: 254, 2187-2190.

Repasky EA, Granger BL, Lazarides E. Widespread occurrence of avian spectrin in nonerythroid cells. Cell 1982: 29, 821-833.

Saxena N, Peterson R., Saxena NK, Russell LD. Microfilaments appear in boar spermatozoa during capacitation in vitro. J exptl Zool 1986: 239, 423-427.

Serres C, Feneux D, Jouannet P. Abnormal distribution of the periaxonemal structures in a human sperm flagellar dyskinesia. Cell Motil Cytosk 1986: 6, 68-76.

Talbot P, Kleve MG. Hamster sperm cross reacts with antiactin. J exptl Zool 1978: 204, 131-136.

Tsukita S, Tsukita S, Ishikawa H, Kurokawa M, Morimoto K, Sobue K, Kakiuchi S. Binding sites of calmodulin and actin on the brain spectrin, calspectin. J Cell Biol 1983: 97, 574-578.

Vera JC, Brito M, Zuvic T, Burzio LO. Polypeptide composition of rat sperm outer dense fibers. A simple procedure to isolate the fibrillar complex. J Biol Chem 1984: 3, 1-9.

Virtanen I, Badley RA, Paasivuo R, Lehto VP. Distinct cytoskeletal domains revealed in sperm cells. J Cell Biol 1984a: 99, 1083-1091.

Virtanen I, Lehto VP, Kallajoki M, Blose SH. Differential localization of α- and β-tubulin in human sperm cells. J Cell Biol 1984b: 99, 41a.

Welch JE, O'Rand MG. Identification and distribution of actin in spermatogenic cells and spermatozoa of the rabbit. Developm Biol 1985: 109, 411-417.

Post-testicular development of mammalian spermatozoa

M.C. ORGEBIN-CRIST

Center for Reproductive Biology Research, Vanderbilt University, Nashville, Tennessee, U.S.A.

INTRODUCTION

After leaving the testis, mammalian spermatozoa undergo a series of changes in the epididymis where they are exposed to epididymal secretions, at ejaculation where they are exposed to the secretions of the various male accessory glands, and in the female genital tract. There the final stages of sperm preparation for fertilization involve capacitation and acrosome reaction, but epididymal sperm maturation may be considered an integral part of the process which permits gamete fusion and formation of the zygote.

1. Development of Sperm Fertilizing Ability

The acquisition of fertilizing ability by epididymal spermatozoa is a gradual process. In all species studied, spermatozoa from the proximal caput cannot fertilize. The fertilizing ability of spermatozoa first become recognizable in the corpus or the proximal cauda, depending on the species [Table 1]. Some individual variations may be observed between animals of the same species, but these variations do not alter drastically the epididymal fertility profile which is characteristic of each species [16].

Direct and reliable information on the extent of fertilizing ability of human spermatozoa at different epididymal levels is lacking. Human caput spermatozoa are unable to bind and penetrate zona-free hamster eggs, while cauda spermatozoa interact readily with such eggs [13, 14]. Although the corpus contains at least some spermatozoa capable of fusion with zona-free hamster eggs, the number of spermatozoa attached to each egg is much lower than the number of cauda or ejaculated spermatozoa [14]. The percentage of spermatozoa exhibiting forward progression also increases as they pass through the epididymis [14]. It appears, therefore, that the human spermatozoon undergoes a process of maturation in the epididymis and becomes functionally competent during epididymal transit. Nonetheless, pregnancies have been reported after vasoepididymostomy which bypasses most of the epididymis [15]. This would imply that spermatozoa need only to be exposed to the environment of the caput to become fertile. However, the quality of sperm motility and the chances of pregnancy increase when the anastomosis is lower [16]. Although sperm maturation in man differs from sperm maturation in animals, both in

155

TABLE 1

REGIONS OF THE EPIDIDYMIS WHERE SPERMATOZOA
ACQUIRE THE CAPACITY TO FERTILIZE

Distal Caput	Marmoset	[1]
Distal Corpus	Rabbit	[2] [3][4]
	Boar	[5]
	Ram	[6]
Proximal Cauda	Mouse	[7] [8]
	Rat	[9] [10]
	Hamster	[11] [12] [1]
	Man	[13] [14]

its timing, which is much shorter [17], and in a greater flexibility as to the epididymal site, it is fair to assume that in man, as in animals, the epididymis makes some critical contributions without which spermatozoa cannot develop forward motility and the ability to fertilize.

In order to fertilize, spermatozoa must ascend the female genital tract, undergo capacitation, recognize and penetrate the zona pellucida, and fuse with the vitelline membrane. The inability of immature spermatozoa to fertilize may be related to one or several of these steps in the fertilization process. In vitro fertilization, which bypasses the need for sperm ascent through the female genital tract, does not improve the fertilizing ability of rabbit caput and corpus spermatozoa [18] and may even delay the process in the mouse [7]. Use of zona-free hamster eggs, which bypasses the need for zona pellucida recognition and penetration, shifts the fertility profile of mouse epididymal sperm to a more proximal region, but does not bring full fertilizing ability to spermatozoa from the distal caput [7]. This suggests that the lower fertility observed with corpus spermatozoa may be mostly due to their inability to undergo acrosome reaction, zona pellucida recognition, and/or zona penetration. Injection of isolated hamster sperm nuclei from the cauda epididymidis directly into the egg cytoplasm permits egg activation and swelling of the nuclei, but caput spermatozoa fail to decondense into pronuclei, although, interestingly, testicular spermatozoa do, suggesting that the resistance to decondensation is a post-testicular acquisition [19]. These experiments emphasize that sperm maturation represents a number of concomitant and successive changes which a normal spermatozoa must undergo before being able to fertilize an egg.

Further maturational changes occur, because spermatozoa which have just gained the ability to fertilize are less likely to induce normal development. Although one study reported no difference in cleavage rate or embryonic mortality after insemination with young epididymal or ejaculated rabbit spermatozoa [20], several other reports have indicated that both in the rabbit and the ram, ova

156

fertilized by immature spermatozoa tend to be delayed in their development, present structural abnormalities, and are less likely to produce viable offspring [4, 2-25]. The maturation of the sperm nuclei in the epididymis is progressive, as indicated by DNA-Feulgen reactivity [26], actinomycin D binding [27-28], and S-S links [29-30]. Although immature epididymal spermatozoa are able to initiate the first steps of fertilization, they are not yet fully competent to induce the development of normal embryos.

Development of Sperm Binding Ability

To fertilize an egg successfully, the spermatozoon must first bind to and penetrate the zona pellucida. Uncapacitated rat testicular spermatozoa, and mouse or hamster caput spermatozoa fail to adhere to the zona pellucida when incubated with rat or mouse eggs, respectively, while spermatozoa from the cauda epididymidis interact readily with the zona surface [31, 32]. The inability of testicular or caput spermatozoa to attach to the zona is not primarily due to their lack of motility because cauda spermatozoa immobilized by cold or lanthanum bind to the zona pellucida [31, 32]. It appears that the inability of immature spermatozoa to interact with the zona pellucida is caused by a deficiency in the adhesive properties of the sperm membrane. The ability to bind to the zona is progressively attained as rat [33] or hamster [34] spermatozoa pass through the corpus epididymidis, and it parallels the development of fertilizing ability which is attained in the rat in the proximal cauda of the epididymis [10].

The physiological significance of this in vitro zona binding is uncertain. In some species, uncapacitated spermatozoa interact with the zona as readily as capacitated spermatozoa. Moreover, spermatozoa from some species (hamster, rat, mouse, rabbit) interact with the zona pellucidae from other species. Human and guinea pig spermatozoa, on the other hand, have a highly specific zona recognition, suggesting that spermatozoa can associate with the zona in a specific as well as in a non-specific manner [35]. The sperm components with which the spermatozoon attaches to the zona surface is also a matter of debate. The plasma membrane over the intact acrosome, the plasma membrane after fusing with the underlying outer acrosomal membrane during the acrosome reaction, the acrosomal matrix exposed by membrane vesiculation, or the inner acrosomal membrane have all been proposed as the structures responsible for sperm attachment to the zona pellucida [35]. In the hamster during in vivo fertilization, it is the remnants of the acrosomal cap loosened as spermatozoa pass through the cumulus which bind to the zona surface. Although some acrosome reacted spermatozoa can bind to the zona using either the inner acrosomal membrane or the plasma membrane over the equatorial segment of the acrosome, most spermatozoa bind to the zona by the fused and vesiculated plasma and outer acrosomal membrane held together by the acrosomal matrix [36]. Whether sperm zona binding observed in the experiments reported above represents specific or non-specific binding, or involves the same sperm components responsible for sperm recognition and attachment to the zona pellucida in vivo at the site of fertilization, has not been resolved. The results do show, however, that during post-testicular maturation modifications of the sperm plasma membrane occur which permit sperm-zona interaction.

Changes in the Sperm Membrane

Interaction of the fertilizing spermatozoon with the egg involves a series of membrane-mediated events: acrosome reaction, release of acrosomal content, penetration through the zona pellucida, contact and fusion with the vitelline membrane. It is likely that modification of the sperm surface during epididymal transit is an important facet of the maturation process.

Structure

Two major classes of membrane proteins have been defined. Integral proteins are embedded within the lipid bilayer and may be exposed at the cytoplasmic face of the membrane, at the external face, or both. The peripheral proteins are present only at the inner or outer face of the membrane and are bound via electrostatic or protein-protein interactions. A particular feature of the sperm membranes (periacrosomal plasma membrane, the underlying outer acrosomal membrane, and the post-acrosomal membrane) is that they are a mosaic of structurally and chemically different domains which participate in sequential manner in the membrane mediated events of fertilization. Each subdomain is associated with distinct cytoskeletal assemblies [37].

During maturation changes occur at the cytoplasmic face of the membrane, within the plane of the membrane, and at the external face of the membrane. In the proximal caput of the rat epididymis a flocculent material fills the space between the plasma membrane and the outer acrosomal membrane [38]. This disappears in the more distal segment of the epididymis. Freeze fracture analysis shows an extensive redistribution of intramembranous particles in the plasma membrane covering the acrosome [38]. The particles are randomly distributed in the plasma membrane of spermatozoa from the proximal caput. In spermatozoa from the distal caput and proximal corpus the particles segregate into plaques of various sizes. The plaque arrangement of the particles disappears in spermatozoa from the proximal cauda epididymidis, where particles are distributed randomly. Those changes are seen only in the acrosomal membrane and not in other regions of the rat sperm surface.

A glycocalyx-like material accumulates on the outer surface of the membrane of distal caput spermatozoa [38]. It is distributed in a patchy fashion reminiscent of the plaque-like arrangement of intramembranous particles seen in freeze-fracture preparations. In the proximal cauda this glycocalyx material begins to detach from the sperm surface and is no longer present on the surface of spermatozoa from the distal cauda. The coincident changes in the distribution of glycocalyx and the intramembraneous particles raises the possibility that the binding of components to the outer membrane might change the structural organization within the membrane. The completion of these membrane changes coincides closely with the acquisition of fertilizing ability in this species [10]. Structural modification of the sperm membrane during epididymal maturation has been demonstrated in other species [39].

Changes occurring at the exterior surface of the sperm plasma membrane have received the most attention because they are likely to reflect contributions of the epididymal environment on which the development of sperm motility and fertilizing ability are dependent. They have been defined in more detail than possible by direct electron microscopic examination, using specific probes which can be visualized by microscopic analysis or by direct biochemical analysis.

Surface Charge

Cell surface differences between immature and mature rabbit epididymal spermatozoa were demonstrated by whole cell electrophoresis; rabbit spermatozoa from the caput have a higher net negative surface charge than spermatozoa from the cauda epididymidis [40]. This observation correlates well with electron microscopic histochemical studies which showed in several species an increased affinity for positively charged colloid particles during passage of spermatozoa through the epididymis and, therefore, an increased density of anionic groups at the sperm surface. The binding is not uniform over the sperm surface, reflecting regional variations in surface character [41-45]. Human spermatozoa show the same increase in anionic sites at the sperm surface during maturation, but the increase is essentially uniform over all regions of the spermatozoon, whereas in other species the negative charges are not equally distributed on the sperm surface [41]. Human spermatozoa fail also to exhibit the preferential orientation in an electrophoretic field characteristic of spermatozoa of many other species [41].

Lectin Binding

Studies with plant lectins have demonstrated changes in the carbohydrate components located at the external face of the spermatozoa during maturation [46-47]. Lectins are plant proteins with different saccharide binding specificities which can be used to identify specific saccharide residues on the surface of mammalian cells. Different methods can be used: 1) radiolabeled lectins to quantitate the accessible binding sites on the sperm surface; 2) ferritin or peroxidase and fluorochrome-conjugated lectins to visualize the binding sites for electron microscopic or light microscopic localization, respectively; and 3) cell agglutinability by specific lectins.

Rabbit caput spermatozoa agglutinates more readily with wheat germ agglutinin (WGA) and Ricinus Communis (RCA) than cauda spermatozoa, but Concanavalin A (Con A) mediated agglutination is identical for caput or cauda spermatozoa [46]. The same results were obtained using ferriting conjugate lectins. Sperm maturation does not seem to alter the pattern of Con A binding sites found on the rabbit sperm surface, but spermatozoa from the caput bound more ferritin-RCA and WGA than did spermatozoa from the cauda [46].

159

Ram caput spermatozoa, like rabbit caput spermatozoa, agglutinate more readily with RCA than cauda spermatozoa, but do not agglutinate with WGA. Cauda spermatozoa agglutinate in the presence of WGA. Con A mediated agglutination is identical for caput and cauda spermatozoa [47].

Both RCA and WGA agglutinate rat caput and cauda spermatozoa, but low concentration of Con A agglutinates only caput spermatozoa. Quantitation of binding sites with radiolabeled Con A indicates that there is a 15-25% decrease in Con A binding by rat spermatozoa as they pass through the epididymis [48]. In addition, there is a difference in the distribution of accessible Con A binding sites over the sperm surface: there is a homogeneous distribution in caput spermatozoa while binding sites are restricted to the head and the junction middle piece-principal piece of cauda spermatozoa [48]. Although the results of lectin binding site localization or lectin mediated agglutination cannot be extrapolated from one species to another, as a whole, they indicate extensive modifications of the sperm surface during maturation. However, they do not reveal if the observed changes result from a loss, masking, binding, or modification of the components.

Composition

Specific macromolecular alterations in the plasma membrane of spermatozoa during maturation have been defined. Intact spermatozoa are incubated with non-penetrating agents which bind or promote the binding of radiolabeled marker molecules to externally located plasma membrane proteins, glycoproteins, or glycolipids. After labeling, the plasma membrane is solubilized and the individually labeled components are identified by affinity chromatography or electrophoretic separation.

Lactoperoxidase catalyzed iodination of spermatozoa from the testis and the cauda and affinity chromatography on concanavalin A sepharose have revealed an increase in concanavalin A binding component on rat spermatozoa during their maturation in the epididymis [49]. Analysis of the labeled components by autoradiography of SDS polyacrylamide gels shows a decreased labeling of high molecular weight components and a major increase in labeling of low molecular weight components during epididymal maturation [48]. Both caput and cauda spermatozoa show similar incorporation into a 37,000 dalton component [48].

Galactose oxidase-^{3}H sodium borohydride treatment used to label membrane glycoproteins possessing oligosaccharide saccharide chains with terminal galactose residues reveals that rat cauda spermatozoa possess a glycoprotein of 37,000 daltons which is not seen in caput spermatozoa [50]. This glycoprotein is first detected in spermatozoa from the proximal cauda [51]. It has been assigned various molecular weights [52-54], but a recent study using electrophoresis in a variety of polyacrylamide gel systems has indicated that its molecular weight may be closer to 24,000 dalton [55]. Treatment with chaotropic agents, high and low salt concentrations and high and low pH,

160

indicates that the molecule is an integral membrane molecule [51, 55-56]. This component is labeled by lactoperoxidase iodination on caput spermatozoa and is modified by glycosylation during sperm maturation in the epididymis [55]. A soluble galactosyltransferase [57] and a lactalbumin-like activity [58-60] are present in rat epididymal fluid. Moreover, caput and cauda rat spermatozoa can incorporate UDP-galactose into a macromolecular fraction and the incorporation is greater when rete testis fluid is present, suggesting that spermatozoa contain macromolecules that are accessible to exogenous and endogenous galactosyltransferases [55, 61-62].

These observations provide evidence that the macromolecular composition of the sperm membrane changes during epididymal transit. It is unlikely that these modifications result from synthesis of new membrane protein by spermatozoa since the machinery necessary for such synthesis is virtually absent from these cells. They are most likely mediated by the epididymal environment.

Epididymal Secretory Proteins

Electrophoretic analyses of epididymal fluid obtained by micropuncture have demonstrated the presence of specific epididymal proteins [63-67]. Some of these specific epididymal proteins have been isolated from epididymal cytosol or fluid and, in some studies, their site of synthesis and their localization have been determined using immunochemical techniques [59, 68-82]. Radiolabel incorporation has also shown that the epididymis synthesizes and secretes proteins [81, 83-85] under the control of androgens [80, 83, 86-90] and that some of these proteins interact with spermatozoa [85, 92, 93]. Protein synthesis and secretion have also been demonstrated by radioautography [94-107].

Protein secretion has been extensively studied in the rat epididymis. A protein complex migrating faster than albumin on non-denaturing polyacrylamide gels was first demonstrated in epididymal fluid [68,70]. The proteins, which were designated B, C, D, E according to their migration in front of albumin, are synthesized by the epididymis and are androgen dependent [86]. Proteins B and C are secreted in the caput, but not in the corpus and the cauda [83, 88], while protein D is synthesized in the caput, corpus, and the cauda and protein E is synthesized only in the corpus and the proximal cauda [108]. Proteins with identical properties have been reported by other investigators: sialoprotein [75], acidic epididymal glycoprotein [70], protein IV [87], 32K [65]. It is likely that these proteins are identical to protein D. Proteins B, C, D, and E have been completely or partially purified [83, 108, 70, 87, 65, 109, 110]. Proteins D and E are glycosylated, but proteins B and C are not [88, 83]. Both proteins B and C display immunological identity as do proteins D and E, since antibodies raised against one protein of the pair react with the other one [83, 108]. Recently, the cDNAs for proteins B and D have been cloned [111]. No variations were found in the nucleotide sequence of several clones for proteins B and C and proteins D and E, suggesting that the primary amino sequence of each pair is identical. Post-translation processing may account for the difference in mobility

between members of each pair. In the case of proteins D and E ·
variations in added carbohydrate side chains is likely to be
responsible [111]. The tissue and species distribution of these
proteins has been investigated by probing Northern blots of RNA with
radioactive cDNA [111, 112]. A weak hybridization was obtained with
the cDNA for proteins D in the rat salivary gland and the mouse
epididymis; otherwise, protein D was not detected in any other rat
tissue or epididymis from guinea pig rabbit, bull, boar, or ram.
Therefore, protein D is specific to the rat epididymis, but a related
protein is present in the epididymis of the mouse.

The association of these androgen dependent secretory proteins
with spermatozoa has been demonstrated with a variety of techniques.
The proteins are present in electrophoretic gels of purified plasma
membrane extracts from cauda but not caput spermatozoa and their
presence on the sperm surface has been demonstrated using
immunolocalization technique [79]. There has been some discrepancies
on the location of proteins D and E on the sperm surface.
Localization has been reported either to be restricted to the
post-acrosomal region [79] or to the acrosome and some part of the
tail [72].

In addition to these androgen-dependent proteins, the rat
epididymis secretes two proteins with a molecular weight of 50,000 and
100,000 daltons [113]. They can be extracted by high ionic strength
media from cauda spermatozoa, but not from caput spermatozoa [113].
Antibodies against either protein react with both proteins, indicating
that they have a high degree of immunological similarity.
Immunolocalization techniques have shown that they are synthesized by
the epithelium of the proximal cauda, that the binding is restricted
to the periacrosomal region of the spermatozoa, and that after in
vitro exposure of corpus spermatozoa to these proteins, binding occurs
to the periacrosomal region as in vivo. Interestingly, there is no
binding after in vitro exposure of caput spermatozoa to these
proteins, suggesting that caput spermatozoa may lack an acceptor site
[113].

Another rat epididymal secretory protein has been reported:
surface radiolabeling of cauda spermatozoa by lactoperoxidase
catalyzed iodination and electrophoretic separation of labeled
polypeptides reveal a major component of 26,000-28,000 daltons, which
is not detected on caput epididymal spermatozoa [48,51]. Protein
separation by two-dimensional polyacrylamide gel electrophoresis and
silver staining reveals a group of polypeptides of differing
isoelectric points in the 26,000-28,000 dalton range which are both
soluble components of epididymal fluid and which are also sperm
associated [66]. Recently, an antibody against the 26,000 component
excised from 2-D gels of purified cauda sperm plasma membranes has
been used for immunolocalization [114]. The monospecific antiserum
stains a single band of 26,000 dalton on western blots of SDS-PAGE
separated plasma membrane from cauda spermatozoa and sperm-free
luminal fluid from the cauda. The antigen is localized exclusively on
the sperm flagellum; it is not present on caput spermatozoa, but first
appeared in the proximal corpus. The antigen present in fluid from

the cauda can bind to caput spermatozoa after in vitro incubation and the localization of the binding is similar to that seen in vivo. This protein is distinct from protein D (or AEG) which is secreted by all levels of the epididymis and is localized on the post-acrosomal region of the spermatozoon. It is distinct also from Proteins B or C, which have an apparent molecular weight of 16,000 and are secreted by the caput epididymidis.

Therefore, in the rat model system which has been the most extensively studied, the epididymis has been shown to secrete several proteins which differ from each other by their size, the regional localization of their secretion and their localization on the sperm surface.

Physiological Significance of Epididymal Secretory Proteins

Numerous studies have shown that sperm maturation depends on the special environment created by the androgen-dependent activity of the epididymal epithelium [for review, 115]. Sperm maturation can be induced in isolated epididymal tubules in vitro by addition of 5α-dihydrotestosterone (5α-DHT) to the culture medium [116]. This stimulatory action of 5α-DHT is suppressed by antiandrogen and inhibitors of RNA and protein synthesis [117]. The development of the sperm fertilizing ability in vitro is, therefore, dependent upon binding of 5α-DHT to receptors, and synthesis of new RNA and protein molecules by the target cells. When the metabolic capacities of the possible target cells are compared, it is likely that the target cells are the epididymal cells rather than the spermatozoa themselves. It was further reported that the stimulatory effect of 5α-DHT upon maturation of epididymal spermatozoa in vitro can be mimicked by the addition of charcoal treated epididymal tissue extracts [118].

The development of zona binding ability is also dependent on androgens. After castration, the zona binding ability of rat spermatozoa is progressively lost from day 3 until day 10, at which time it is nil. Similarly, fertilizing ability declines from day 5 after castration until day 10. Testosterone supplementation maintains both zona binding ability and fertilizing ability at control levels [33].

The conditions required for the development of sperm zona binding ability have been determined using hamster epididymal tubules cultured in vitro [34]. Few spermatozoa from the proximal corpus epididymis cultured over 24 hours without androgens are able to bind to the zona pellucida. A significant increase in zona binding ability is observed when the epididymal tubules are cultured in the presence of 5α-DHT. Cyproterone acetate or cycloheximide significantly decreases sperm zona binding ability when compared with cultures to which only 5α-DHT is added [34]. Thus, it appears that, in the hamster, the development of zona binding ability during epididymal transit is, like the development of fertilizing ability in the rat and the rabbit, under androgen regulation and is dependent on de novo protein synthesis.

When a crude fraction enriched in epididymal glycoproteins and depleted of androgens is added to the cultured epididymal tubule, sperm zona binding ability increases significantly. This increase is observed only with epididymal cytosol from intact animals. Epididymal cytosol from hamster castrated for 30 days fails to elicit an increase in sperm zona binding ability [119]. These results emphasize similarities between the regulation of sperm zona binding ability and fertilizing ability, and stress the importance of the interaction between spermatozoa and epididymal secretory proteins.

However, the function of the several epididymal glycoproteins which have been purified to date has not been completely elucidated. Addition of purified epididymal glycoprotein, presumed to be equivalent to proteins D and E, to immature rat spermatozoa in vitro promotes their ability to attach to the zona [32], while immunization of male rats with the same protein complex causes a decrease in the motility, zona binding ability, and fertilizing ability of their spermatozoa [120]. Furthermore, addition of specific antibodies against proteins D and E to rat spermatozoa in vitro prior to artificial insemination decreases their fertilizing ability [121]. These results suggest that proteins D and E (or AEG) may have a role in sperm zona binding.

Rat epididymal cells in culture synthesize glycosidases in a high molecular weight precursor form and process this forms intracellularly to the mature active enzyme which is secreted in the medium [122]. The function of these secreted glycoprotein modifying enzymes has not been determined, but could be important in modifying the composition and properties of the sperm surface.

Secretion of epididymal proteins having α-lactalbumin modifying activity toward galactosyltransferase has also been reported [58-60]. Like the glycosidases, they could have a potential important function, since the epididymal fluid contains a galactosyltransferase activity [57] and that a membrane-bound galactosyltransferase is thought to be instrumental in mouse sperm-zona binding [123]. It has been suggested that proteins B and C correspond to this α-lactalbumin-like proteins [59]. However, Brooks has shown recently that they do not [112]: A comparison of the amino acid sequence of proteins B and C with the protein sequence database has revealed that they bear no sequence homology with rat α-lactalbumin, but that they belong instead to the α_{2u}-globulin family. The functions of these proteins with α_{2u}-globulin activity, of the 50,000-100,000 dalton component localized on the sperm acrosome, the 26,000 dalton component localized on the sperm tail, or the 24,000 dalton integral protein probably glycosylated during epididymal transit have not yet been determined.

CONCLUSION

Although a considerable body of information is available on various aspects of sperm maturation (motility, zona binding, fertilizing ability, modification of the sperm membrane), few reports have shown a relationship between a specific molecular event and the expression of fertilizing ability. We have recently shown that the infertility of genetically-defined mice bearing a single gene mutation (bouncy) is related to the inability of their spermatozoa to interact with the zona pellucida [124]. The epididymal epithelium from this mutant fails to synthesize and secrete a low molecular weight protein present in the congenic wild type. Mutants of this type may provide us with a tool for dissecting the sequence of events occurring during maturation.

This review is an updated version of reviews which have appeared in proceedings of several meetings: Orgebin-Crist MC. Epididymal physiology and sperm maturation. In: Progress in Reproductive Biology. Epididymis and Fertility: Biology and Pathology, eds. Bollack C, Clavert A, Hubinont PO. Basel, Karger, 1981: 8; 80-89. Orgebin-Crist MC, Olson GE. Epididymal maturation. In: The Male in Farm Animal Reproduction, eds. Courot M. The Netherlands, Martinus Nijhoff, 1984: 80-102. Orgebin-Crist MC. Physiologie de l'epididyme et maturation du sperme: etat actuel des connaissances. Contraception-Fertilite-Sexualite 1986: 14; 487-495.

REFERENCES

1. Moore HDM. An assessment of the fertilizing ability of spermatozoa in the epididymis of the marmoset monkey (Callithrix jacchus). Intl J Androl 1981: 4: 321-320.

2. Nishikawa Y, Waide Y. Studies on the maturation of spermatozoa. I. Mechanism and speed of transition of spermatozoa in the epididymis and their functional changes. Bull natn Inst Agr Sci 1952: Series G; 369-381.

3. Bedford JM. Development of the fertilizing ability of spermatozoa in the epididymis of the rabbit. J Exp Zool 1967: 163; 319-329.

4. Orgebin-Crist MC. Maturation des spermatozoides dans l'epididyme chez le lapin: pouvoir fecondant et mortalite embryonnaire chez des lapines inseminees avec du sperme epididymaire. Annls Biol anim Biochim Biophys 1967: 7; 373-389.

5. Holtz W, Smidt D. The fertilizing capacity of epididymal spermatozoa in the pig. J Reprod Fert 1976: 46; 227-229.

6. Fournier-Delpech S, Colas G, Courot M, Ortavant R. Observations on the motility and fertilizing ability of ram epididymal spermatozoa. Annls Biol anim Biochim Biophys 1977: 17; 987-990.

7. Pavlok A. Development of the penetration activity of mouse epididymal spermatozoa in vivo and in vitro. J Reprod Fert 1974: 36; 203-205.

8. Hoppe PC. Fertilizing ability of mouse sperm from different epididymal regions and after washing and centrifugation. J Exp Zool 1975: 192; 219-222.

9. Blandau RJ, Rumery RE. The relationship of swimming movements of

epididymal spermatozoa to their fertilizing capacity. Fert Steril 164: 15; 571-579.

10. Dyson ALMB, Orgebin-Crist MC. Effects of hypophysectomy, castration and androgen replacement upon the fertilizing ability of rat epididymal spermatozoa. Endocrinology 1973: 93; 391-402.

11. Horan AH, Bedford JM. Development of the fertilizing ability of spermatozoa in the epididymis of the Syrian hamster. J Reprod Fert 1972: 30; 417-423.

12. Cummins JM. Effects of epididymal occlusion on sperm maturation in the hamster. J Expl Zool 1976: 197; 187-190.

13. Hinrichsen MJ, Blaquier JA. Evidence supporting the existence of sperm maturation in the human epididymis. J Reprod Fert 1980: 60; 291-294.

14. Moore HDM, Hartman TD, Pryor JP. Development of the oocyte-penetrating capacity of spermatozoa in the human epididymis. Intl J Androl 1983: 6; 310-318.

15. Schoysman R. Epididymal causes of male fertility. In: Progress in Reproductive Biology, Epididymis and Fertility, Biology and Pathology, eds. Bollack C, Clavert A, Hubinont PO. Basel, Karger, 1981: 8; 102-113.

16. Shoysman R, Bedford JM. The role of the human epididymis in sperm maturation and sperm storage as reflected in the consequences of epididymovasostomy. Fert Steril 1986: 46; 293-299.

17. Rowley MJ, Teshima F, Heller CG. Duration of transit of spermatozoa through the human male ductular system. Fert Steril 170: 21; 390-395.

18. Brackett BG, Hall JL, Oh YK. In vitro fertilizing ability of testicular, epididymal, and ejaculated rabbit spermatozoa. Fert Steril 1978: 29; 571-582.

19. Yanagimachi R. Behavior of nuclei of testicular caput and cauda epididymal spermatozoa injected into hamster eggs. Biol Reprod 177: 16; 315-321.

20. Overstreet JW, Bedford JM. Embryonic mortality in the rabbit is not increased after fertilization by young epididymal spermatozoa. Biol Reprod 1976: 15; 54-57.

21. Orgebin-Crist MC. Maturation of spermatozoa in the rabbit epididymis: Delayed fertilization in does inseminated with epididymal spermatozoa. J Reprod Fert 1968: 16; 29-33.

22. Orgebin-Crist MC. Studies on the function of the epididymis. Biol Reprod 1969: 1; 155-175.

23. Orgebin-Crist MC, Jahad N. Delayed cleavage of rabbit ova after fertilization by young epididymal spermatozoa. Biol Reprod 1977: 16; 358-362.

24. Fournier-Delpech S, Colas G, Courot M, Ortavant R, Brice G. Epididymal sperm maturation in the ram: Motility, fertilizing ability and embryonic survival after uterine artificial insemination in the ewe. Annls Biol anim Biochim Biophys 1979: 19: 597-605.

25. Fournier-Delpech S, Colas G, Courot M. Observations sur les premiers clivages des oeufs intratubaires de brebis apres fecondation avec des spermatozoides epididymaires ou ejacules. C R Acad Sc Paris serie III 1981: t 292; 515-517.

26. Esnault C, Courot M, Ortavant R. Transport and maturation of epididymal spermatozoa in domestic animals. In: The Biology of Spermatozoa, eds. Hafez ESE, Thibault CG, Basel, Karger, 1975: 28-35.

27. Calvin HI, Bedford JM. Stimulation of actinomycin D-binding to eutherian sperm chromatin by reduction of disulphide bonds. J Reprod Fert 1974: 36; 225-229.

28. Darzynkiewica Z, Gledhill BL, Ringertz NR. Changes in deoxyribonucleoprotein during spermiogenesis in the bull. [^{3}H]-actinomycin D binding capacity. Expl Cell Res 1969: 58; 435-438.

29. Bedford JM, Calvin HI. The occurrence and possible functional significance of -S-S- crosslinks in sperm heads, with particular reference to Eutherian mammals. J Exp Zool 1974: 188; 137-156.

30. Calvin HI, Bedford JM. Formation of disulphide bonds in the nucleus and accessory structures of mammalian spermatozoa during maturation in the epididymis. J Reprod Fert 1971: Suppl. 13; 65-75.

31. Saling PM. Development of the ability to bind to zonae pellucidae during epididymal maturation: Reversible immobilization of mouse spermatozoa by lanthanum. Biol Reprod 1982: 26; 429-436.

32. Orgebin-Crist MC, Fournier-Delpech S. Sperm-egg interaction: Evidence for maturational changes during epididymal transit. J Androl 1982: 3; 429-433.

33. Fournier-Delpech S, Hamamah S, Anthony CT, Courot M, Orgebin-Crist MC. Hormonal regulation of zona binding ability and fertilizing ability of rat epididymal spermatozoa. Gam Res 1984: 9; 21-30.

34. Cuasnicu PS, Gonzaez Echeverria F, Piazza A, Blaquier JA. Addition of androgens to cultured hamster epididymis increases zona recognition by immature spermatozoa. J Reprod Fert 1984: 70; 541-547.

35.	Yanagimachi R. Mechanisms of fertilization in mammals. In: Fertilization and Embryonic Development In Vitro, eds. Mastroianni L Jr, Biggers JD. New York, Plenum Press, 1981: Chapter 5; 82-182.

36.	Yanagimachi R, Phillips DM. The status of acrosomal caps of hamster spermatozoa immediately before fertilization in vivo. Gam Res 1984: 9; 1-19.

37.	Olson GE, Winfrey VP, Flaherty SP. Cytoskeletal assemblies of mammalian spermatozoa. In: Cell Biology of the Testis and Epididymis. eds. Orgebin-Crist MC, Danzo BJ. New York, Ann N Y Acad Sci, 1987: in press.

38.	Suzuki F, Nagano T. Epididymal maturation of rat spermatozoa studied by thin sectioning and freeze-fracture. Biol Reprod 1980: 22; 1219-1231.

39.	Olson GE. Changes on the sperm surface during maturation in the reproductive tract. In: Ultrastructure of Reproduction, eds. van Blerkom J, Motta P, The Netherlands, Martinus Nijoff B.V., 1984: 95-108.

40.	Bedford JM. Changes in the electrophoretic properties of rabbit spermatozoa during passage through the epididymis. Nature 1964: 200; 1178-1180.

41.	Beford JM, Calvin H, Cooper GW. The maturation of spermatozoa in the human epididymis. J Reprod Fert 1973: Suppl. 18; 199-213.

42.	Cooper GW, Bedford JM. Acquisition of surface charge by the plasma membrane of mammalian spermatozoa during epididymal maturation. Anat Rec 1971: 169; 300-301.

43.	Yanagimachi R, Noda YD, Fujimoto M, Nicolson GL. The distribution of negative surface charges on mammalian spermatozoa. Am J Anat 1972: 135; 497-520.

44.	Flechon JE. Ultrastructural and cytochemical modifications of rabbit spermatozoa during epididymal transport. In: The Biology of Spermatozoa. Transport, Survival, and Fertilizing Ability, eds. Hafez ESE, Thibault CG. Basel, Karger, 1975: 36-45.

45.	Courtens JL, Fournier-Delpech S. Modifications in the plasma membrane of epididymal ram spermatozoa during maturation and incubation in utero. J Ultrastruc Res 1979: 68; 136-148.

46.	Nicholson GL, Usui N, Yanagimachi R, Yanagimachi H, Smith JR. Lectin-binding sites on the plasma membranes of rabbit spermatozoa. Changes in surface receptors during epididymal maturation and after ejaculation. J Cell Biol 1977: 74; 950-962.

47.	Hammerstedt RH, Hay SR, Amann RP. Modification of ram sperm membranes during epididymal transit. Biol Reprod 1982: 27; 745-755.

48. Olson GE, Danzo BJ. Surface changes in rat spermatozoa during epididymal maturation. Biol Reprod 1981: 24; 431-443.

49. Fournier-Delpech S, Danzo BJ, Orgebin-Crist MC. Extraction of concanavalin A affinity material from rat testicular and epididymal spermatozoa. Ann Biol anim Bioch Biophys 1977: 17; 207-213.

50. Olson GE, Hamilton DW. Characterization of the surface glycoproteins of rat spermatozoa. Biol Reprod 1978: 19; 26-35.

51. Olson GE, Orgebin-Crist MC. Sperm surface changes during epididymal maturation. In: The Cell Biology of the Testis, eds. Bardin CW, Sherins RJ. New York, Ann N Y Acad Sci, 1982: 383; 372-391.

52. Jones R, Pholpamool C, Setchell BP, Brown CR. Labelling of membrane glycoproteins on rat spermatozoa collected from different regions of the epididymis. Bioch J 1981: 200; 457-460.

53. Brown CR, von Gloss KI, Jones R. Changes in plasma membrane glycoproteins of rat spermatozoa during maturation in the epididymis. J Cell Biol 1983: 96; 256-264.

54. Zeheb R, Orr GA. Characterization of a maturation-associated glycoprotein on the plasma membrane of rat caudal epididymal sperm. J Biol Chem 1984: 259; 839-848.

55. Hamilton DW, Wenstrom JC, Baker JB. Membrane glycoproteins from spermatozoa: Partial characterization of an integral $Mr = \simeq$ 24,000 molecule from rat spermatozoa that is glycosylated during epididymal maturation. Biol Reprod 1986: 34; 925-936.

56. Moore A, Ensrud K, Baker J, Wenstrom J, Hamilton DW. Preliminary observations on a second $Mr = \simeq$ 24,000 membrane molecule from rat spermatozoa. In: Cell Biology of the Testis and Epididymis, eds. Orgebin-Crist MC, Danzo BJ. New York, Ann N Y Acad Sci, 1987: in press.

57. Hamilton DW. UDP-galactose: N-acetyl-glucosamine galactosyl-transferase in fluids from rat rete testis and epididymis. Biol Reprod 1980: 23; 377-385.

58. Hamilton DW. Evidence for α-lactalbumin-like activity in reproductive tract fluids of the male rat. Biol Reprod 1981: 25; 385-392.

59. Jones R, Brown CR. Association of epididymal secretory proteins showing α-lactalbumin-like activity with plasma membrane of rat spermatozoa. Biochem J 1982: 206; 161-164.

60. Byers SW, Qasba PK, Paulson HL, Dym M. Immunocytochemical localization of α-lactalbumin in the male reproductive tract. Biol Reprod 1984: 30; 171-178.

61. Hamilton DW, Gould RP. Galactosyltransferase activity associated with rat epididymal spermatozoan maturation. Anat Rec 1980: 196; 71a.

62. Hamilton DW, Gould RP. Preliminary observations on enzymatic galactosylation of glycoproteins on the surface of rat caput epididymal spermatozoa. Intl J Androl 1982: Suppl. 5; 73-80.

63. Koskimies AA, Kormano M. Proteins in fluids from different segments of the rat epididymis. J Reprod Fert 1975: 43; 345-348.

64. Turner TT, Plesums JL, Cabot CL. Luminal fluid proteins of the male reproductive tract. Biol Reprod 1979: 21; 883-890.

65. Wong PYD, Tsang AYF. Studies on the binding of a 32 K rat epididymal protein to rat epididymal spermatozoa. Biol Reprod 1982: 27; 1239-1246.

66. Olson GE, Hinton BJ. Regional differences in luminal fluid polypeptides of the rat testis and epididymis revealed by two-dimensional gel electrophoresis. J Androl 1985: 6; 20-34.

67. Flickinger CJ, Herr JC, Ertl KE. Identification and isolation of epididymal luminal proteins of the mouse. J Androl 1986: 7; 163-168.

68. Fournier S. Electrophorese des proteines du tractus genital du rat. I. Presence dans le sperme epididymaire d'une glycoproteine migrant vers l'anode a pH = 8.45. C R Soc Biol 1968: 162; 568-571.

69. Fournier-Delpech S, Bayard F, Boulard C. Isolement, extraction and caracterisation d'une sialoproteine du sperme epididymaire du rat par electrophorese sur polyacrylamide. C R Soc Biol 1973: 167; 543-546.

70. Lea OA, Petrusz P, French FS. Purification and localization of an acidic epididymal glycoprotein (AEG): A sperm coating protein secreted by the rat epididymis. Intl J Anat 1978: Suppl. 2; 592-607.

71. Garberi JC, Kohane AC, Cameo MS, Blaquier JA. Isolation and characterization of specific rat epididymal proteins. Mol Cell Endocrinol 1979: 13; 73-82.

72. Kohane AC, Garberi JC, Cameo MC, Blaquier JA. Quantitative determination of specific proteins in the rat epididymis. J Steroid Biochem 1979: 11; 671-674.

73. Kohane AC, Cameo MS, Piniero L, Garberi JC, Blaquier JA. Distribution and site of production of specific proteins in the rat epididymis. Biol Reprod 1980a: 23; 181-187.

74. Kohane AC, Gonzalez Echeverria FM, Piniero L, Blaquier JA.

Interaction of proteins of epididymal origin with spermatozoa. Biol Reprod 1980b: 23; 737-742.

75. Faye JC, Duguet L, Mazzuca M, Bayard F. Purification, radioimmunoassay, and immunohistochemical localization of a glycoprotein produced by the rat epididymis. Biol Reprod 1980: 23; 423-432.

76. Moore HDM. Localization of specific glycoproteins secreted by the rabbit and hamster epididymis. Biol Reprod 1980: 22; 705-718.

77. Moore HDM. Effects of castration on specific glycoprotein secretions of the epididymis in the rabbit and hamster. J Reprod Fert 1981: 61; 347-354.

78. Wong PYD, Tsang AYF, Lee WM. Origin of the luminal fluid proteins of the rat epididymis. Intl J Androl 1981: 4; 331-341.

79. Brooks DE, Tiver K. Location of epididymal secretory proteins on rat spermatozoa. J Reprod Fert 1983: 69; 651-657.

80. Kohane AC, Piniero L, Blaquier JA. Androgen controlled synthesis of specific proteins in the rat epididymis. Endocrinology 1983: 112; 1590-1596.

81. Thomas TS, Reynolds ALB, Oliphant G. Evaluation of the site of synthesis of rabbit sperm acrosome stabilizing factor using immunocytochemical and metabolic labelling techniques. Biol Reprod 1984: 30; 693-705.

82. Rifkin JM, Olson GE. Characterization of maturation dependent extrinsic proteins of the rat sperm surface. J Cell Biol 1985: 100; 1582-1591.

83. Brooks DE. Secretion of proteins and glycoproteins by the rat epididymis: Regional differences, androgen dependence and effects of protease inhibitors, procaine and tunicamycin. Biol Reprod 1981: 25; 1099-1117.

84. Sylvester SR, Skinner MW, Griswold MK. A sulphated glycoprotein synthesized by Sertoli cells and by epididymal cells is a component of the sperm membrane. Biol Reprod 1986: 31; 1087-1101.

85. Klinefelter GR, Hamilton DW. Synthesis and secretion of proteins by perifused caput epididymal tubules and association of secreted proteins with spermatozoa. Biol Reprod 1985: 33; 1017-1027.

86. Cameo MS, Blaquier JA. Androgen controlled specific proteins in the rat epididymis. J Endocrinol 1976: 69; 47-55.

87. Jones R, Brown CR, Von Glos KE, Parker MG. Hormonal regulation of protein synthesis in the rat epididymis. Characterization of androgen dependent and testicular fluid dependent proteins. Biochem J 1980: 188; 667-676.

88. Brooks ˙DE, Higgins SJ. Characterization and androgen dependence of proteins associated with luminal fluid and spermatozoa in the rat epididymis. J Reprod Fert 1980: 59; 363-375.

89. Jones R, Von Glas KI, Brown CR. Characterization of hormonally regulated secretory protein from the caput epididymis of the rabbit. Biochem J 1981: 195; 105-114.

90. Jones R, Fournier-Delpech S, Willadsens SA. Identification of androgen dependent proteins synthesized in vitro by the ram epididymis. Reprod. Nutr. Develop. 1982: 22; 495-504.

91. Brooks DE. Effects of androgens on protein synthesis and secretion in various segments of the rat epididymis as analyzed by two dimensional gel electrophoresis. Mol Cell Endocrinol 1983: 29; 255-270.

92. Voglmayr JK, Fairbanks G, Vespa DB, Colella JR. Studies on the mechanism of surface modification in ram spermatozoa during the final stages of differentiation. Biol Reprod 1982: 26; 483-500.

93. Brooks DE. Selective binding of specific rat epididymal secretory proteins to spermatozoa and erythrocytes. Gam Res 1983: 7; 367-376.

94. Neutra M, Leblond CP. Radiographic comparison of the uptake of galactose-H^3 in the Golgi region of various cells secreting glycoproteins or mucopolysaccharides. J Cell Biol 1966: 30; 137-150.

95. Vendrely C, Durliat M. Etude autoradiographique de l'influence de l'hormone male sur l'incorporation de leucine dans le tractus genital de la souris blanche. Archiv Anat Histol Embryol 1968: 51; 735-740.

96. Kopecny V. Epididymal luminal contents labelling after ^{14}C or ^{3}H-lysine administration in the mouse. Acta Histochem Bd 1971: 40; 116-122.

97. Bennett G, Leblond CP, Haddad A. Migration of glycoprotein from the Golgi apparatus to the surface of various cell types as shown by radioautography after labeled fucose injection in rats. J Cell Biol 1974: 60; 258-284.

98. Kanka J, Kopecny V. An autoradiographic study of macromolecular synthesis in the ductus epididymis of the mouse. I. DNA, RNA and protein. Biol Reprod 1977: 16; 421-427.

99. Kopecny V, Pech V. An autoradiographic study of macromolecular syntheses in the epithelium of the ductus epididymis in the mouse II. Incorporation of L-fucose-1-^{3}H. Histochemistry 1977: 50; 229-238.

100. Flickinger CJ. Synthesis, transport and secretion of protein in

the initial segment of the mouse epididymis as studied by electron microscope radioautography. Biol Reprod 1979: 20; 1015-1030.

101. Flickinger CJ. Regional differences in synthesis, intracellular transport and secretion of protein in the mouse epididymis. Biol Reprod 1981: 25; 871-883.

102. Flickinger CJ. Synthesis and secretion of glycoprotein by the epididymal epithelium. J Androl 1983: 4; 157-161.

103. Flickinger CJ. Radioautographic analysis of the secretory pathway for glycoproteins in the principal cells of the mouse epididymis exposed to [^{3}H] fucose. Biol Reprod 1985: 32; 377-389.

104. Flickinger CJ, Wilson KM, Gray HD. The secretory pathway in the mouse epididymis as shown by electron microscopy of principal cells exposed to monensin. Anat Rec 1984: 210; 435-448.

105. Orgebin-Crist MC, Menezo Y. A continuous flow method for organ culture of rabbit epididymis: Morphology, amino acid utilization, glucose uptake, RNA and protein synthesis. J Androl 1980: 1; 289-298.

106. Fain-Maurel MA, Dadoune JP, Jauzein-Leau F. Protein secretion by the principal cells of the mouse epididymis evidenced by in vitro incorporation of tritiated leucine. Reprod Nutr Develop 1983: 23; 175-182.

107. Orgebin-Crist MC, Hoffman LH, Olson GE, Skudlarek MD. Secretion of proteins and glycoproteins by perifused rabbit corpus epididymal tubules: Effect of castration. Am J Anat 1987: in press.

108. Brooks DE. Purification of rat epididymal proteins 'D' and 'E,' demonstration of shared immunological determinants and identification of regional synthesis and secretion. Int J Androl 1982: 5; 513-524.

109. Bayard F, Dugvet L, Mazzuca M, Faye JC. Study of glycoprotein produced by the rat epididymis. In: Reproductive Processes and Contraception, ed. McKerns KW. New York, Plenum Press, 1981: 393-405.

110. Garberi JC, Fontana JD, Blaquier JA. Carbohydrate composition of specific rat epididymal protein. Int J Androl 1982: 5; 619-626.

111. Brooks DE, Means AR, Wright EJ, Singh SP, Tiver KK. Molecular cloning of the cDNA for two major androgen-dependent secretory proteins of 18.5 kilodaltons synthesized by the rat epididymis. J Biol Chem 1986: 261; 4956-4961.

112. Brooks DE. Androgen regulated epididymal secretory proteins associated with post-testicular sperm development. In: Cell

Biology of the Testis and Epididymis, eds. Orgebin-Crist MC, Danzo BJ. New York, Ann N Y Acad Sci, 1987: in press.

113. Rifkin JM, Olson GE. Characterization of maturation dependent extrinsic proteins of the rat sperm surface. J Cell Biol 1985: 100; 1587-1591.

114. Olson GE, Lifsics MR, Winfrey VP, Rifkin JM. Modification of the rat sperm flagellar plasma membrane during maturation in the epididymis. J Androl 1987: in press.

115. Orgebin-Crist MC, Danzo BJ, Davies J. Endocrine control of the development and maintenance of sperm fertilizing ability in the epididymis. In: Handbook of Physiology-Endocrinology V, eds. Greep RO, Hamilton DW. Baltimore, Williams & Wilkins, 1975: Chapter 15; 319-338.

116. Orgebin-Crist MC, Jahad N, Hoffman LH. The effects of testosterone, 5α-dihydrotestosterone, 3α-androstanediol, and 3α-androstanediol on the maturation of rabbit epididymal spermatozoa in organ culture. Cell Tiss Res 1976: 167; 515-525.

117. Orgebin-Crist MC, Jahad N. The maturation of rabbit epididymal spermatozoa in organ culture: inhibition by antiandrogens and inhibitors of RNA and protein synthesis. Endocrinology 1978: 103; 46-53.

118. Orgebin-Crist MC, Jahad N. The maturation of rabbit epididymal spermatozoa in organ culture: Stimulation by epididymal cytoplasmic extracts. Biol Reprod 1979: 21; 511-515.

119. Gonzalez-Echeverria F, Cuasnicu PS, Piazza AD, Pineiro L, Blaquier JA. Addition of an androgen-free epididymal protein extract increases the ability of immature hamster spermatozoa to fertilize in vivo and in vitro. J Reprod Fert 1984: 71; 433-437.

120. Fournier-Delpech S, Courot M, Dubois MP. Decreased fertility and motility of spermatozoa from rats immunized with a prealbumin epididymal-specific glycoprotein. J Androl 1985: 6; 246-250.

121. Cuasnicu PS, Gonzalez-Echeverria F, Piazza AD, Cameo MS, Blaquier JA. Antibodies against epididymal glycoproteins block fertilizing ability in the rat. J Reprod Fert 1984: 72; 467-471.

122. Skudlarek MD, Orgebin-Crist MC. Glycosidases in cultured rat epididymal cells: Enzyme activity, synthesis and secretion. Biol Reprod 1986: 35; 167-178.

123. Lopez LC, Bayna EM, Litoff D, Shaper NL, Shaper JH, Shur BD. Receptor function of mouse sperm surface galactosyltransferase during fertilization. J Cell Biol 1985: 101; 1501-1510.

124. Holland MK, Orgebin-Crist MC. Epididymal protein synthesis and secretion in strains of mice bearing single gene mutations which affect fertility. Biol Reprod 1987: in press.

174

Fertilizing capacity of human spermatozoa

R. YANAGIMACHI

Department of Anatomy and Reproductive Biology, University of Hawaii School of Medicine, Honolulu, Hawaii, U.S.A.

Clinicians working in infertility clinics wish to know if the spermatozoa of the man in a particular infertile couple have full fertilzing ability or functional defects. Clinical investigators working on the development of fertility-controlling agents wish to know whether the agents they are testing have improved or eliminated the fertilizing capacity of spermatozoa. In both cases,they will first perform a conventional semen analysis. This relatively simple test provides us with much infomration about the potential fertility of spermatozoa. If the analysis reveals that the man in question does not ejaculate spermatozoa at all, it is then obvious that the man is infertile. If the man ejaculates only a very few actively motile spermatozoa with normal morphology, he is most likely to be subfertile or infertile. However, clinicians and clinical investigators have experienced that some of the so-called oligospermic men are capable of impregnating their partners from time to time. Alternatively, there are some men who are clinically infertile in spite of the fact that their semen (sperm) parameters are all apparently "normal." The cervical mucus penetration test has proven to be very valuable in detecting some sperm dysfunctions as well as immunological problems. Can we accurately assess the fertilzing capacity of human spermatozoa by examining the spermatozoa in semen and/or examining the behavior of spermatozoa in the cervical mucus? I do not think so, because sperm's most important function, that is fertilizing eggs, cannot be evaluated by examining the spermatozoa far away from their target, the eggs. The most important question to be answered is whether or not the spermatozoa can actually fertilize eggs. I do not mean to imply that the conventional semen analysis and the cervical mucus penetration test are meaningless. On the contrary, they are invaluable in assessing various, important features of spermatozoa. What I wish to say is that we would be able to assess the fertility of spermatozoa more accurately by performing not only the

175

conventional tests, but also an assay which directly evaluates the fertilzing capacity of spermatozoa. No single test or assay can evaluate every parameter of sperm function. Various tests, each evalu ating a specific feature(s) or function(s) of spermatozoa, should be combined to assess as accurately as possible the fertility of spermatozoa.

Currently available Techniques to assess the Fertilizing Ability of Human Spermatozoa
The most certain way of evaluating the fertilizing ability of human spermatozoa is to deposit them in the genital tract of a woman of proven fertility during the periovulatory period and later determine the success or failure of fertilization and/or pregnancy. If we could inseminate a group of females of proven fertility (as we do with laboratory and farm animals), we would certainly know whether or not the spermatozoa in question are fertile. This approach is impractical for obvious ethical reasons. An alternate approach would be to inseminate living human eggs in vitro. This again would provoke an ethical debate unless spermatozoa and eggs were obtained from married couples directly. Routine use of living human eggs for any biological assay is not legally permitted in most countries.

The next best way of assessing fertilizing capacity of human spermatozoa would be to inseminate non-viable human eggs in vitro. Overstreet and Hembree (1) collected immature eggs from ovaries, kept them frozen (2), then inseminated them in vitro after thawing. The ability of spermatozoa to cross the zona pellucida was used as an indicator of fertilizing capacity of spermatozoa. This apparoach is sensible, since zona penetration by spermatozoa requires capacitation, the acrosome reaction and strong motility of spermatozoa, all of which are essential for normal fertilization. The spermatozoa which are able to penetrate through the zona pellucida are also most likely to be able to fertilize the egg. It is very unlikely that the eggs Overstreet and Hembree inseminated were alive because the freeze-thawing technique these investigators employed was not specifically designed to keep the eggs alive. Even if some frozen-thawed eggs did remain alive, it is very improbable that fertilized immature eggs could develop into live fetuses.

Yanagimachi et al. (3) proposed to use non-living human eggs in assessing fertilizing capacity of human spermatozoa. Instead of freezing eggs, we placed the eggs in a highly concentrated solution of neutral salt (e.g., 2 M ammonum sulfate or 1.5 M magnesium chloride). Upon transfer into such a solution, the egg proper (vitellus) shrank, leaving the zona pellucida apparently intact. The eggs were kept in the solution for up to several months in a refrigerator (4°C). When washed thoroughly and inseminated in vitro with human spermatozoa, eggs were

176

surrounded by many spermatozoa and sperm penetration into the perivitelline space through the zona was observed (Fig. 1). The eggs to be stored in salt solutions do not need to be fully mature. However, we routinely cultured ovarian eggs in vitro to allow their "maturation." The cultured eggs were freed from surrounding follicle cells before they were transferred to salt solutions. Since we published the original paper mentioned above, I realized that the pH of the salt solution must be kept at, or near, neutral. Otherwise, the zonae may be swollen or even disssolved after a prolonged storage. One of the salt solutions I currently use is 2 M ammonium sulfate in 20 mM Tris-buffered (pH 7.0-7.2) supplemented with 0.5% dextran (200-300 K). Recently, Dr. Alex Lopata's laboratory in Australia sent me, via a regular air mail, a number of salt-stored human eggs. They were the eggs not "fertilized" by IVF. When washed thoroughly and inseminated in vitro in my laboratory, zonae pellucidae were penetrated by spermatozoa. Some of the eggs had dead (killed by ammonium sulfate solution) spermatozoa in the zona pellucida, and in some instances in the perivitelline space. I could differentiate these spermatozoa from the spermatozoa I had added by labeling the latter with either FITC (4) or monobromobimane (5). Zonae pellucidae of some of the eggs I received from Australia were not penetrated by spermatozoa I added. It is possible that these eggs had been actually fertilized by IVF, but were judged by Dr. Lopata's group as "unfertilized" (probably because they did not cleave). Since the zona pellucida of human egg becomes rejective to spermatozoa soon after fertilization (cf. 6), it is advisable not to use fertilized eggs in evaluating the sperm's ability to cross the zona pellucida. According to the result obtained by Bedford (7), gibbon's eggs may be used as substitutes for human eggs. Zonae pellucidae of this primate, unlike those of other species, permit penetration by human spermatozoa. Although gibbon's eggs might be the ideal substitutes for human eggs, gibbons are unfortunately an endangered species and their eggs are not readily accessible to most investigators.

Another example of the substitutes for human eggs is golden hamster eggs (8). Zona-intact hamster eggs cannot be penetrated by human spermatozoa, however, once zonae are removed, zona-free eggs can be penetrated by human spermatozoa, provided the spermatozoa have been capacitated and acrosome-reacted (Fig. 2). Uncapacitated, acrosome-intact spermatozoa are unable to penetrate zona-free eggs. In other words, zona-free hamster eggs can be used to assess whether the human spermatozoa under investigation have the ability to undergo both capacitation and the acrosome reaction. Testing the fertilzing capacity of human spermatozoa using zona-free hamster eggs has been called the hamster egg penetration test or simply the hamster test (9).

There has been considerable debate about the validity of this test (for discussion, see 9-13). It is my impression that investigators who have had extensive experience with this test claim the existence of a close (but, not absolute) correlation between the result of hamster test and the fertility of spermatozoa, whereas, those who have engaged the test less extensively tend to be opposed to this view. What the hamster test reveals is the sperm's ability to undergo a "spontaneous" acrosome reaction and fuse with the egg plasma membrane. If a particular sperm sample contains many spermatozoa that are capable of undergoing the spontaneous acrosome reaction, high egg penetration rates are expected. It is important to note that there is considerable variation among men in relation to the degree and speed of sperm capacitation (and of spontaneous acrosome reaction) in media. The composition of the medium also greatly influences the degree and speed of capacitation and of the acrosome reaction (for discussion, see 9). If a negative result (no or very poor egg penetration) is obtained in the first test, experiments should be repeated by changing the condition of sperm incubation.

Positive results in the hamster test (high egg penetration rates) must be interpreted cautiously. If the spermatozoa retain strong motility for many hours and penetrate the vast majority of zona-free hamster eggs, the chance of these spermatozoa to fertitilize human eggs is high. However, there is no absolute guarantee that the spermatozoa yielding high egg penetration rates in the hamster test can fertilize human eggs in vivo. Spermatozoa in vivo must "negotiate" a variety of "barriers" in the female genital tract before reaching the eggs. The hamster test can not predict whether or not the spermatozoa would be able to overcome these barriers. If the spermatozoa become weakly motile during incubation, they (even if they are able to penetrate zona-free hamster eggs) may not be able to fertilize zona-intact human eggs because zona penetration by spermatozoa requires not only the acrosome reaction, but also strong motility of spermatozoa.

Negative results in the hamster test (no egg penetration at all or very low penetration rates, say less than 10%) appear to indicate the dysfunction of the spermatozoa with some accuracy. The failure of sperm penetration in zona-free hamster eggs indicates that either the spermatozoa could not survive long enough or could not undergo the acrosome reaction efficiently. As stated already, we must repeat experiments when a negative result is obtained in the first trial. The sperm preincubation time could have been too short or too long, or the condition of the medium could have been suboptimal. If we consistently obtain negative results after repeated tests under various conditions, while spermatozoa from men of proven fertility (controls) always give highly positive results, then "abnormalities" of the

178

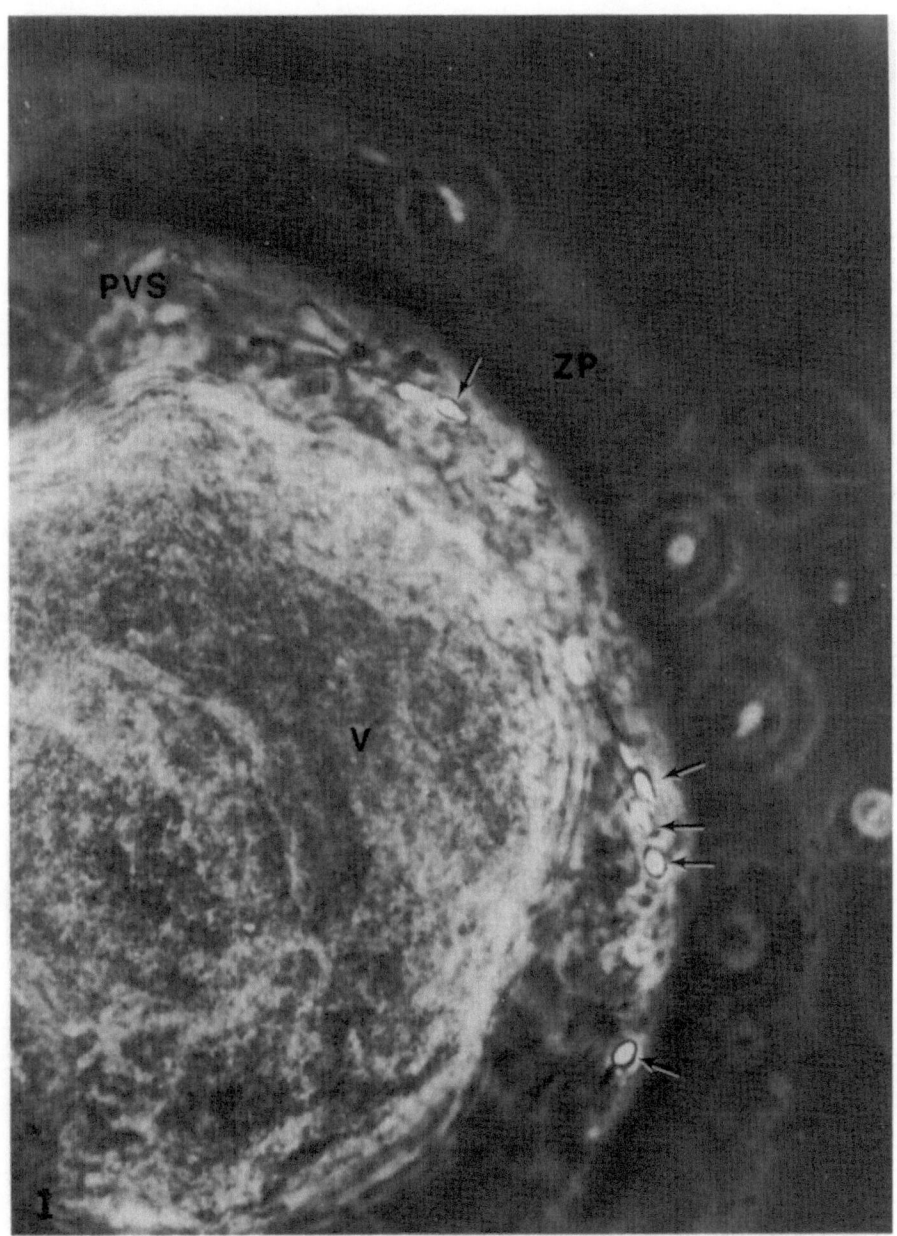

Fig. 1. Sperm penetration through the zona pellucida of a salt-stored human egg. Heads of several spermatozoa (arrows) which have passed through the zona pellucida (ZP) are seen in the perivitelline space (PVS). The vitellus (V) is shrunk and no sperm incorporation is observed. X 1,500.

spermatozoa must be suspected.

Zona-free hamster eggs are not human eggs. A substitute must be recognized as a substitute. The most prominent difference between zona-intact human eggs and zona-free hamster eggs is, obviously, the presence or absence of the zona pellucida. Both the zona pellucida and the cumulus oophorus surrounding the zona appears to have the ability to promote or induce the acrosome reaction of human spermatozoa (14, 15). In other words, intact human eggs have "built-in" acrosome reaction-promoting or -triggering mechanisms. Zona-free hamster eggs do not. If we incorporate some acrosome reaction-promoting (or -inducing) agents into the sperm incubation medium, zona-free hamster eggs could be penetrated by human spermatozoa at higher rates than in a medium without them. In other words, the sensitivity of the hamster test would be increased by such a manipulation. In fact, according to Aitken et al. (16), the sensitivity of the hamster test is increased significantly by treating human spermatozoa with a calcium ionophore which accelerates (or induces) the acrosome reaction. Mortimer (17) and Mortimer et al. (18) reported that the sensitivity of the hamster test is increased considerably when human spermatozoa are preincubated in a sperm-capacitating medium containing strontium in stead of calcium. Perhaps, higher proportions of the spermatozoa undergo their acrosome reactions after such a treatment. Before we decide whether the hamster test should be adopted or abandoned as a means of assessing the fertility of human spermatozoa, we should, at least, confirm the claims by these investigators.

As stated previously, the hamster test evaluates the ability of human spermatozoa to undergo the acrosome reaction and to fuse with the egg plasma membrane. Detecting the acrosome reaction in human spermatozoa, without time-consuming electron microscopy, was extremely difficult or very imprecise. It is now possible by staining the contents of the acrosomal caps (19-22). It would be worthwhile to investigate whether or not we can assess the fertilizing capacity of human spermatozoa by examining the ability of spermatozoa to undergo the acrosome reaction (23). Detecting the acrosome reaction in human spermatozoa is now relatively simple. However, as the ability of spermatozoa to undergo the acrosome reaction is just one of many abilities the spermatozoa must have to effect fertilization, we cannot expect from examining the acrosome reaction alone, we can predict the fertility of spermatozoa with great accuracy. Only when combined with other tests evaluating other functional parameters of spermatozoa, would this test be very valuable.

180

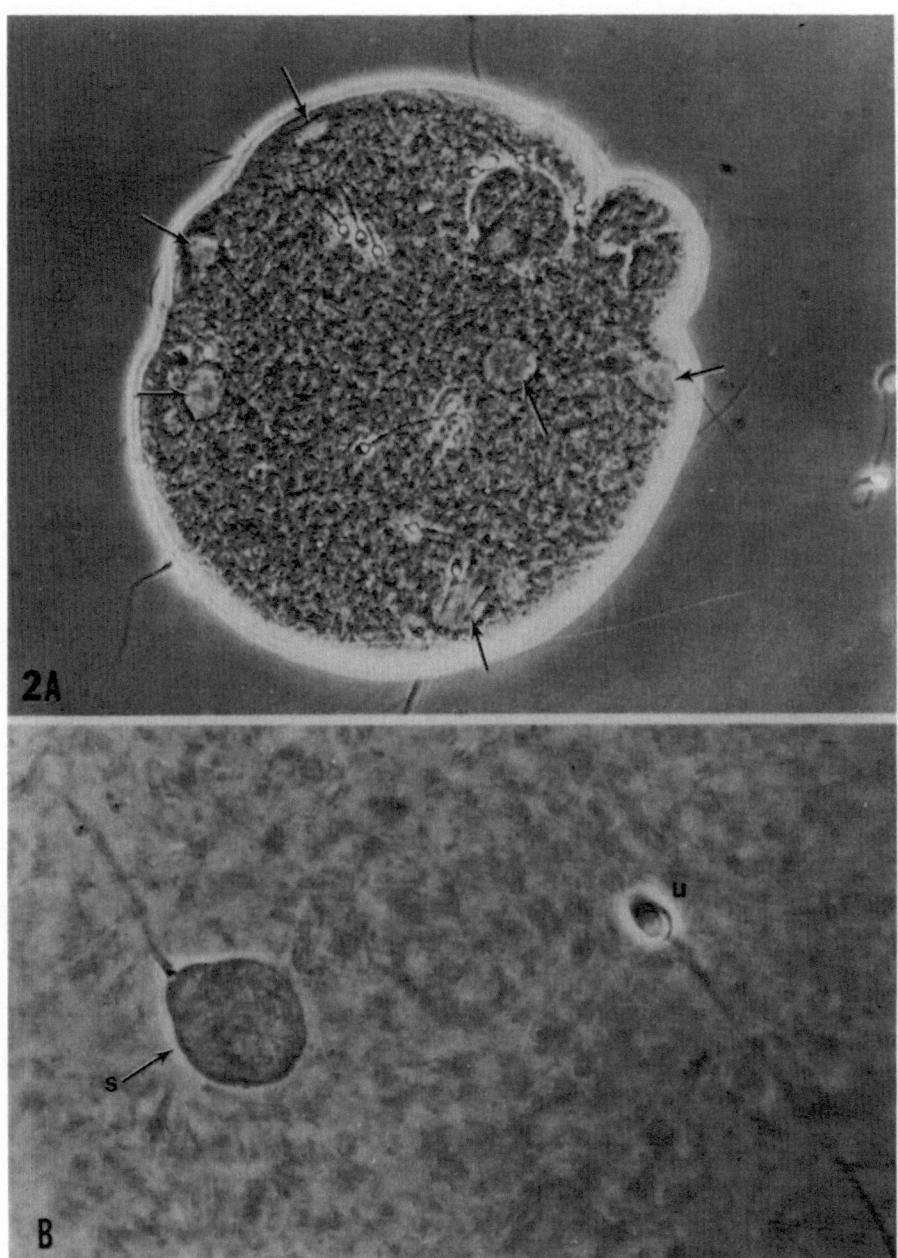

Fig. 2. Zona-free hamster eggs penetrated by human
spermatozoa. Arrows indicate swollen sperm heads in the
vitellus. Compare the size of swollen sperm head (s) with
that of unswollen head (u). A, a living specimen,
compressed between a slide and coverslip, X 600. B, a
fixed and stained specimen, X 2,200.

Is the Fertility and Infertility of Spermatozoa absolute?

Spermatozoa of oligospermic men are usually unable to fertilize the eggs in vivo. This does not necessarily mean that ologspermic men do not produce fertile spermatozoa. There may be thousands of potentially fertile spermatozoa in the ejaculate, but the chance of these spermatozoa fertilizing eggs in vivo is slim because the female genital tract does not allow all the ejaculated spermatozoa to reach the ampullary region of the oviduct where fertilization takes place. Even if hundreds of millions of spermatozoa are ejaculated into the vagina, less than 1,000 or only 10^{-5} % of the total are allowed to reach the ampulla (24). If we bypass the female tract and allow the spermatozoa of oliospermic men to meet the eggs directly (by IVF procedure), they may be capable of fertilizing eggs in some cases (e.g., 25). In other words, spermatozoa which are "infertile" in vivo, could be fertile in vitro.

The Spermatozoa of men with Kartagner's syndrome are completely immotile because of the lack of dynein arms (26). Since strong motility of spermatozoa is essential for successful sperm passage through certain regions of the female tract (e.g., the cervix filled with cervical mucus) and the egg investments (particularly the zona pellucida), immotile spermatozoa have no chance to fertilize the eggs not only in vivo, but also in IVF. Are these spermatozoa completely infertile? Maybe not. According to Aitken et al. (27), these spermatozoa can be induced to undergo the acrosome reaction. Acrosome-reacted spermatozoa, although they are still immotile, can fuse with zona-free hamster eggs. Therefore, if one or a few of the acrosome-reacted spermatozoa are placed microsurgically within the perivitelline space of an egg, the egg might be fertilized successsfully. This is an extreme example, but the spermatozoa which are unable to fertilize eggs by the ordinary IVF technique may not necessarily be infertile.

Some men may be infertile because of defective epididymides. Some of them may not be able to ejaculate spermatozoa at all, or even if they can, their semen may contain only immature spermatozoa. Is their situation hopeless? Maybe not. If there is an indication that the epididymis contains mature spermatozoa, they can be collected from the epididymis and used for IVF (28). If the caput epididymis or the testis contains spermatozoa, these spermatozoa can be collected. Although such spermatozoa will not be able to fertilize eggs even by IVF, they may have potential to "fertilize" eggs. According to Uehara and Yanagimachi (29), the nuclei of hamster epididymal and testicular spermatozoa can develop into apparently normal sperm (male) pronuclei when injected microsurgically into the cytoplasm of mature eggs. Although we did not prove that the eggs with

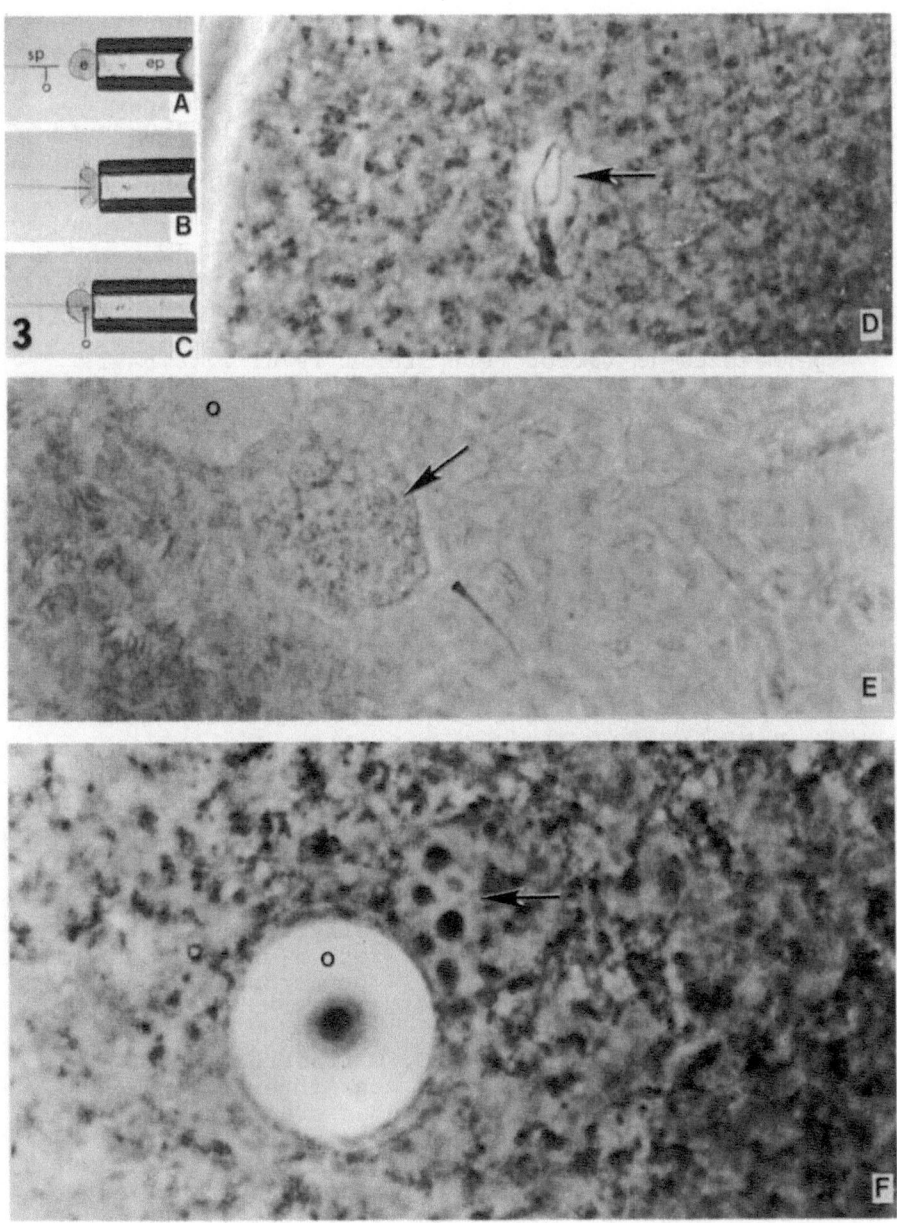

Fig. 3. Development of pronuclei from the heads of freeze-dried human spermatozoa. A-C, microsurgical injection of a spermatozoon into a hamster egg. D, a sperm head (arrow) soon after injection. E-F, developing sperm pronuclei (arrows). The egg of E was fixed and stained before photographed. Abbreviations: e, hamster egg; eg, egg-holding pipette; o, paraffin oil to be injected or injected into eggs; sp, sperm-injecting pipette.

183

injected sperm nuclei can develop into live young, Markert (30) has reported that mouse eggs injected with nuclei of mature mouse spermatozoa can develop to blastocysts. There is no reason to believe that, given the proper technique, an egg with an injected sperm nucleus cannot develop into a normal fetus. Even the nuclei of freeze-dried spermatozoa may be able to participate in the development of eggs (31) (Fig. 3).

References

1) Overstreet JW, Hembree WC. Penetration of the zona pellucida of nonliving human oocytes by human spermatozoa in vitro. Fert Steril 1976: 27;815-31.

2) Overstreet JW, Yanagimachi R, Katz DF, Hayashi K, Hanson FW. Penetration of human spermatozoa into the human zona pellucida and the zona-free hamster egg: A study of fertile donors and infertile patients. Fert Steril 1980: 33;534-42.

3) Yanagimachi R, Lopata A, Odom CB, Bronson RA, Mahi CA, Nicolson GL. Retention of biologic characteristics of the zona pellucida in highly concentrated salt solution: The use of salt-stored eggs for assessing the fertilizing capacity of spermatozoa. Fertil Steril 1980: 31;562-74.

4) Blazak WF, Overstreet JW, Katz DF, Hanson FW. A comparative in vitro assay of human sperm fertilizing ability utilizing contrasting fluorescent sperm markers. J Androl 1982: 3;165-71.

5) Cummins JM, Fleming AD, Crozet N, Kuehl TJ, Kosower N, Yanagimachi R. Labelling of living mammalian spermatozoa by the fluorescent alkylating agent, monobromobimane (MB). Immobilization upon exposure to ultraviolet light and analysis of acrosome reaction. J Exp Zool 1986: 237;375-382.

6) Sathananthan AH, Trounson AO. Ultrastructure of cortical granule release and zona interaction in monospermic and polyspermic human ova fertilized in vitro. Gamete Res 1982: 6;225-34.

7) Bedford JM. Sperm/egg interaction: the specificity of human spermatozoa. Anat Rec 1977: 188;477-88.

8) Yanagimachi R, Yanagimchi H, Rogers BJ. The use of zona-free animal ova as a test-system for the assessment of fertilizing capacity of human spermatozoa. Biol Reprod 1976: 15;471-76.

9) Yanagimachi R. Zona-free hamster eggs: their use in assessing fertilizing capacity and examining chromsomes of human spermatozoa. Gamete Res 1984 :10; 363-73.

10) Wolf DP, Sokoloski JE, Quigley MM. Correlation of human in vitro fertilization with the hamster egg bioassay. Fert Steril 1983: 40;53-59.

11) Foreman R, Cohen J, Fehilly CB, Fishel SB, Edwards RG. The application of the zona-free hamster egg test for the prognosis of human in vitro fertilization. J In Vitro Fert 1984: 1;166-71.

12) Rogers BJ. The sperm penetration assay: its usefulness reevaluated. Fert Steril 1985: 43; 821-40.

13) Aitken RJ. Diagnostic value of the zona-free hamster oocyte penetration test and sperm movement characteristics in oligozoospermia. Int J Androl 1984: 8;348-56.

14) Tesarik J. Comaprision of acrosome-inducing activities of human cumulus oophorus, follicular fluid and ionophore A23187 in human sperm populations of proven fertilizing ability in vitro. J Reprod Fertil 1985: 74;383-88.

15) Singer SL, Lambert H, Overstreet JW, Hanson FW, Yanagimachi R. The kinetics of human sperm binding to the human zona pellucida and zona-free hamster oocyte in vitro. Gamete Res 1985: 12;29-39.

16) Aitken RJ, Ross A, Hargreave T, Richardson D, Best F. Analysis of human sperm function following exposure to the ionophore A23187. J Androl 1984: 5;321-29.

17) Mortimer D. Comparision iof the fertilizing ability of human spermatozoa preincubated in calcium- and strontium-co ntaining media. J Exp Zool 1986: 237; 21-24.

18) Mortimer D, Curtis EF, Dravland JE. The use of strontium-substituted media for capacitating human spermatozoa: An improved sperm preparation method for the zona-freee hamster egg penetration test. Fertil Steril 1986: 46;97-103.

19) Talbot P, Chacon RS. A triple-stain technique for evaluating normal acrosome reacosome of human spermatozoa. J Exp Zool 1981: 215;201-8.

20) Talbot P, Dudenhausen E.   Factors affecting triple
    staining of human spermatozoa.   Stain Tech 1981: 56;
    307-9.

21) Wolf DP, Boldt J, Byrd W, Bechtol KB.   Acrosomal
    status evaluation in human ejaculate sperm with
    monoclonal antibodies. Biol Reprod 1985; 32;
    1157-62

22) Cross NL, Morales P, Overstreet JW.   Two simple
    methods for detecting sperm acrosome reaction.   J
    Androl 1986: 7;28P.

23) Topfer-Petersen E, Heissler E, Schill WB.   The
    kinetic of acrosome reaction: an additional sperm
    parameter? Andrologia 1985: 17;224-27.

24) Mortimer D.   Sperm transport in the human female
    reproductive tract. Oxford Rev Reprod Boil 1983: 5;
    30-36.

25) Cohen J, Edwars R, Fehilly C, Fishel S, Hewitt J,
    Pyrdy J, Rowland G, Steptoe P, Webster J.   In vitro
    fertilization: a treatment for male infertility.
    Fert Steril 1985: 43;422-32.

26) Afzelius BA, Eliasson R, Johnson O, Lindholmer C.
    Lack of dynein arms in immotile human spermatozoa. J
    Cell Biol 1975: 66;225-232.

27) Aitken RJ, Ross A, Lees MM.   Analysis of sperm
    function in Kartagner's syndrome.   Fert Steril
    1983: 40;696-98.

28) Temple-Smith PD, Southwick GJ, Yates CA, Trounson A,
    De Kretser DM.   Human pregnancy by in vitro
    fertilization (IVF) using sperm aspirated from the
    epididymis.   J In Vitro Fert Embryo Trans 1985: 2;
    119-122.

29) Uehara T, Yanagimachi R.   Behavior of nuclei of
    testicular, caput and cauda epididymal spermatozoa
    injected into hamster egg.   Biol Reprod 1977: 16;
    315-21.

30) Markert CL. Fertilization of mammalian eggs by sperm
    injection.   J Exp Zool 1983: 228;195-201.

31) Uehara T, Yanagimachi R.   Microsurgical injection of
    spermatozoa into hamster eggs with subsequent
    transformation of sperm nuclei into male pronuclei.
    Biol Reprod 1976: 15;467-470.

Factors modulating immunocompatibility of spermatozoa: role of transglutaminase and SV-IV, one of the major proteins secreted from the rat seminal vesicle epithelium

S. METAFORA, G. PELUSO, G. RAVAGNAN[1],
V. GENTILE[2], A. FUSCO[2], R. PORTA[2]

*CNR Institute of Protein Biochemistry and Enzymology, Naples, Italy;
[1]CNR Institute of Experimental Medicine, Rome, Italy; [2]Institute of
Chemistry and Biological Chemistry, First Medical School, University
of Naples, Naples, Italy*

INTRODUCTION

In the process of mammal sexual reproduction two distinct, highly specialized cells, the spermatozoon and the egg, originated from two genetically different individuals of the same species, come in contact in an appropriate environment, recognize each other and then fuse in the fertilization process to generate a multipotent developing zygote. This type of reproduction system is, however, potentially hazardous for two main reasons: 1) the evolution has worked out in the vertebrate species an extremely efficient immunosurveillance system to protect the precious "self" of the single individuals against the possible intrusion of molecular or cellular foreign "non-self" invaders; 2) the male gamete is antigenic. In fact, it has been demonstrated by a variety of conventional immunological techniques that the epididymal spermatozoa and their immediate precursors accomodate on their cell surface a cargo of "non-self" specific, highly immunogenic autoantigens, alloantigens (including those of the major histocompatibility system) as well as xenoantigens (1, 2). Meanwhile the latter are irrelevant to an immune response to a normal sexual mating, both autoantigens and alloantigens possess the ability to elicit a dangerous immune response in the female, whose genital tract is, in general, fully capable of both cellular and humoral immune response to "non-self", alien invading elements. The "non-selfness" of the epididymal spermatozoa is most probably due to the fact that spermatogenesis takes place only long after the tolerance susceptible phase of mammalian development is terminated.

To avoid sensitization of adult males to differentiation "non-self" autoantigens and to other potential autoantigens, as well as to ensure the survival of sperm cells in the female hostile immunological environment, the evolution has developed specific structural and biochemical defensive mechanisms. Structural protective devices occurring in testis are the tight junctions of the Sertoli cells which line the seminiferous tubules to mechanically prevent both the contact between immunocompetent cells and differentiating spermatozoa, and the passage of circulating antibodies into the tubular lumen (blood-testis barrier)(1). Moreover, it has been reported the presence in the secretion of <u>in vitro</u> cultured Sertoli cells of specific

-proteins (3) that can possibly play a role in a putative biochemical system involved in the immunoprotection of the differentiating gametes in this district.

In the normal epididymis there are, as a consequence of a leakage in the blood-testis barrier, many immunocompetent cells (T and B lymphocytes, macrophages, etc.) potentially able to attack and destroy the maturing autoantigenic sperm cells (1). In physiological conditions, however, no immunological attack is launched here, probably because of the following three reasons: 1) presence of a high proportion of suppressor T cells within the epididymal epithelium (1, 2); 2) immunosuppressive activity of the immature germinal cells (4); 3) action of soluble immunosuppressive protein factors secreted from the epithelial cells lining the epididymis duct lumen (unpublished data).

The surface antigenic characteristics of the epididymal spermatozoa change following their mixing in the ejaculate with the secretions of the male accessory sexual glands. By interacting with proteins and other factors of seminal plasma, the sperm cells lose their immunological reactivity becoming adapted to avoid rejection by the immunocompetent elements present in both seminal fluid and female genital tract (2). In our laboratory we have found that the numerous T lymphocytes, occurring in normal human seminal fluid (5), are quiescent, not activable by the common mitogens and unable to function as responders in mixed lymphocyte cultures (see below).

The factors and the molecular mechanism(s) responsible for the specific state of tolerance of the immunocompetent cells present in the reproductive tract of both sexes are still unknown. Unclear is also the biochemical mechanism underlying the reduced immunogenicity of the spermatozoa following ejaculation. Several seminal plasma components (proteins such as uteroglobin (UG), transglutaminase (TGase), different nucleases and proteases; other factors like polyamines, prostaglandins, endorphins, etc.) have been reported to have powerful immunomodulatory activities, leading to a marked impairment of the specific functions of most cells of the immune system (T cells, B cells, NK cells, macrophages, etc.) (2). Particularly interesting in this context is the UG/TGase system, present in the rabbit. It has been reported that UG (a small MW secretory protein (Mr = 15,000) consisting of two identical subunits synthesized under steroid hormone control in the seminal vesicle, prostate, uterus and lung of rabbits) is transformed by TGase (a Ca^{2+} dependent enzyme, present in large amount in the rat coagulating gland secretion (6) and detectable also in human seminal plasma (7)) in a modified form of higher MW able to suppress epididymal sperm antigenicity by binding to the sperm surface (8, 9).

It has been recently demonstrated, by comparative computer analysis, that UG fragment has a strong structural homology (70%; see Ref. 10) with a peptide present in the protein SV-IV (rat seminal vesicle no 4, according to its electrophoretic mobility in SDS-PAGE) whose synthesis in the vesicular epithelium is under strict transcriptional testosterone control (11, 12, 13). The immunological analysis by polyclonal antisera, specific for SV-IV, has also shown a marked cross-reaction between the two proteins (10). SV-IV-related antigens have been detected in other rat organs including lung, liver, brain, uterus (10) and in human seminal fluid (14). It is worth noiting that the UG ability to reduce dramatically, in the presence of TGase and calcium ions, the immunogenicity of the rabbit epididymal spermatozoa may well be crucial in the prevention of the immune response to the sperm deposited in the female genital tract during coitus.

On the basis of these data and by considering the marked homology between UG and SV-IV, we have investigated to verify whether the vesicular protein SV-IV has, in the presence of TGase and calcium ions, the same ability as UG to reduce the epididymal sperm immunogenicity. In addition, experiments were performed to ascertain whether SV-IV possesses immunosuppressive properties to explain the absence of immunoreactivity of the T lymphocytes present in the human seminal fluid.

MATERIALS AND METHODS

Purification of proteins and TGase assay

The protein SV-IV was purified to homogeneity from adult rat (Fisher-Wistar strain) seminal vesicle secretion according to Ostrowski et al. (11). (^{125}I) SV-IV was prepared by using the Chloramine T method (15).

Pure TGase was prepared from guinea pig liver according to Connellan et al. (16).

The radiometric assay used to determine TGase activity has been previously described (17).

UG, purified from rabbit uterus, was kindly supplied by Dr. T. Tancredi (CNR, Arco Felice, Naples).

Isolation of cells

Human peripheral blood lymphocytes (PBL) were obtained from normal individuals by Ficoll-Paque fractionation. Splenocytes were isolated from adult male rats (Fisher-Wistar strain) by a modified standard method (18).

Sodium dodecyl sulphate (SDS) 15%-polyacrylamide gel electrophoresis (PAGE).

The SDS-PAGE analysis of the (^{125}I) SV-IV bound to lymphocytes was performed on 15% acrylamide gels (acrylamide: N, N^1-methylenebisacrylamide = 15%: 0.13%) containing 0.1% SDS with Tris-glycine as running buffer (19). The protein samples, dissolved in O'Farrell sample buffer (20), were boiled for 3 min before loading on gel. Fluorography of the dried gels was performed according to the method of Bonner & Laskey (21).

Protein concentration measurement

The protein content of the analyzed samples was measured by the technique of Lowry et al. (22) or by absorbance at 280 nm.

RESULTS AND DISCUSSION

TGase-modified form(s) of the protein SV-IV possess the ability to bind **in vitro** to the rat epididymal sperm surface.

As already mentioned in the Introduction, it has been demonstrated that a TGase-modified form of UG is able to suppress in vitro the epididymal sperm antigenicity in rabbit (8) and that a marked structural homology exists between the rabbit UG and the protein SV-IV (10). Enzymological studies carried out by our group pointed to SV-IV as an effective substrate for guinea pig liver TGase in both polyamination and polymerization reactions (17 and unpublished data). We have extended these studies by investigating the interaction of SV-IV with rat epididymal spermatozoa and the role of TGase in this event (17). Our data indicate that, although TGase does not seem to produce ε (γ-glutamyl)lysine bonds between SV-IV and sperm surface proteins, SV-IV is able to bind to the epididymal sperm surface only after its pretreatment with TGase and Ca^{2+}. Therefore, modified forms of SV-IV possessing the specific ability to bind to the epididymal sperm surfaces are generated by TGase. SDS-PAGE analysis of sperm-bound SV-IV have shown that only the high molecular weight form(s) of the protein were able to bind to the sperm surface, thus suggesting for TGase a role as modulator of SV-IV binding properties (17).

189

TGase-modified form(s) of SV-IV dramatically reduces the epididymal sperm immunogenicity.

To study the effect of the SV-IV binding on the sperm immunogenicity, we used mixed cultures of rat splenocytes and epididymal spermatozoa taken from heterologous males. The splenocytes were used as responder cells actively incorporating labelled thymidine when stimulated by the alloantigens of the epididymal spermatozoa. The latter were subjected to various treatments to ascertain the effects of SV-IV and TGase in masking sperm antigenicity. From the data shown in Table I it is clear

TABLE I

Suppression of epididymal sperm immunogenicity by uteroglobin (UG), rat seminal vesicle no IV (SV-IV) and transglutaminase (TGase) as measured by [^{3}H]thymidine incorporation into splenocytes.

TREATMENT OF SPERM	c.p.m. $\pm$ S.E.M.	Percent of control
RABBIT (see Reference no 8)		
None	11,500 $\pm$ 350	100
UG	6,700 $\pm$ 200	58
TGase	10,800 $\pm$ 182	94
UG plus TGase	760 $\pm$ 80	7
RAT*		
None	7,860 $\pm$ 680	100
SV-IV	4,300 $\pm$ 510	55
TGase	4,700 $\pm$ 520	60
SV-IV plus TGase	1,900 $\pm$ 280	24

* Rat (Fisher strain) epididymal spermatozoa (1x10^5 cells) were treated, for 2 h at 37°C, in 0.2 ml of HBSS (Hanks balanced salt solution) modified by addition of 2.5 mM CaCl$_2$ and 5 mM dithiothreitol (DTT), either with 8 µg of SV-IV in the presence or absence of 10 µg of TGase purified from guinea pig liver (16). The sperm cells were then washed three times with 10 ml of unmodified HBSS and finally mixed with 4x10^5 rat (Brown-Norway strain) splenocytes. Complete HBSS was used to culture the splenocyte-sperm mixture. Each culture was done in triplicate. After 4.5 days of incubation at 37°C in a CO$_2$ incubator, 1 µCi of [^{3}H] thymidine (specific activity, 2 Ci/mmol, Amersham) was added and the culture continued for another 12 hours. The cells were then harvested, washed, filtered on Millipore filters and counted in a Beckman scintillation counter after Instagel addition to the dried filters. The viability of the rat splenocytes was assessed by Con A stimulation and [^{3}H] thymidine incorporation (splenocytes only, 200 $\pm$ 75 cpm; splenocytes plus Con A, 28,000 $\pm$ cpm).

that the treatment of the sperm cells with TGase and SV-IV leads to a marked decrease of sperm immunogenicity. Similar results were previously obtained by Mukherjee et al. (8) with UG in the rabbit system. The masking of sperm antigenicity could be, in our opinion, of paramount importance for the prevention of a dangerous immune response to the sperm cells by the female genital tract during coitus. The treatment with TGase alone was apparently effective in reducing the antigenicity of epididymal spermatozoa only in rats. This discrepancy is most probably related to differences between the two systems (two TGases of different origin act on different protein substrates occurring on the surfaces of epididymal spermatozoa obtained from different species). It is worth noting that in both systems the intensity of the masking effect of the enzyme plus reactive proteins (SV-IV or UG) was not the sum of the single effects of the reagents. This result could be explained by taking into account that the reactive proteins in the presence of TGase are modified molecular forms of the native species. Fig. 1 shows the stimulation of labelled thymidine incorporation in rat (Brown-Norway) splenocytes by different number of rat (Fisher) epididymal spermatozoa. This experiment was done to find in the rat system the best spermatozoa/splenocytes ratio for the mixed culture. As a final comment, it is interesting to observe that among different cultures used (spermatozoa (Spz) Fisher + Splenocytes (Spl) Fisher; Spz Brown + Spl Brown; Spz Fisher + Spl Brown; Spz Brown + Spl Fisher) the most active in thymidine incorporation was the culture Spz Fisher + Spl Brown, the less active being the Spz Fisher + Spl Fisher culture.

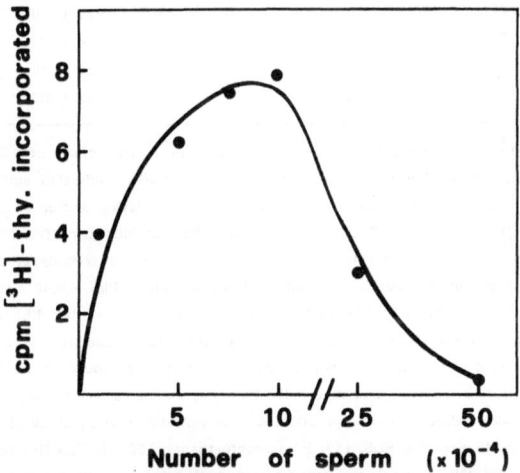

Fig. 1 - Stimulation of (^{3}H) thymidine incorporation in rat (Brown-Norway strain) splenocytes by different number of rat (Fisher strain) epididymal spermatozoa. The mixed cell cultures were performed according to the procedure reported in Table I.

The T lymphocytes occurring in the normal human seminal fluid are quiescent not activated cells.

It has been recently reported (5) the presence in normal human ejaculate of numerous T lymphocytes. Their analysis by fluorescent specific monoclonal antibodies and fluorescence activated cell sorter has demonstrated that these seminal elements are quiescent cells, not activated (Ia absent), not activable with the common mitogens, unable to act as responders in mixed cultures and apparently endowed with unexpected properties of immaturity (presence on their surface of T_6, T_9, T_{10} antigens). Why are these lymphocytes so unreactive to such common non-self stimuli? The answer to this question is, most probably, to avoid immunological damage of the mature elements of the germ line. Such unreactivity could be explained by the binding of some immunosuppressive factors, contained in the human seminal plasma, to the lymphocytes. By using polyclonal antisera specific to SV-IV our group has recently reported the presence in human seminal fluid of a family of proteins immunologically related to SV-IV both in soluble and sperm-bound form (14). At this point a new question arises: are SV-IV in rat and UG in rabbit, whose homologues seem to be present in humans, able to act as immunosuppressive factors? In other words, are the SV-IV or UG-immunorelated forms, occurring in human seminal plasma

191

endowed with immunomodulatory properties responsible for seminal T lymphocyte unreactivity?

TGase-modified form(s) of SV-IV possess the ability to bind to the human PBL surfaces.

Since SV-IV is an effective TGase substrate (17) and since an active TGase was shown to be present in the human seminal plasma (7), we performed experiments to measure the binding of radioactive SV-IV to human PBL in the presence or absence

TABLE II

Effect of purified guinea pig liver TGase on the $[^{125}I]$SV-IV binding to human peripheral blood lymphocytes (PBL).

Composition of the incubation mixture	$\begin{bmatrix} 125_I \end{bmatrix}$ SV-IV bound to PBL (cpm $\pm$ S.E.M.)
PBL + $[\,^{125}I\,]$ SV-IV	815 $\pm$ 90
PBL + $[\,^{125}I\,]$ SV-IV + Ca^{2+}	3,280 $\pm$ 315
PBL + $[\,^{125}I\,]$ SV-IV + TGase	800 $\pm$ 85
PBL + $[\,^{125}I\,]$ SV-IV + TGase + Ca^{2+}	6,052 $\pm$ 600

The $[^{125}I]$ SV-IV binding assay mixture contained, in a final volume of 0.3 ml, 3×10^6 PBL, 50 mM Tris-HCl buffer (pH 7.5), 120 mM NaCl, 5 mM DTT and 1.25 µg $[^{125}I]$ SV-IV (5×10^6 cpm). Where indicated 2.5 mM CaCl$_2$, and/or 10 µg of purified guinea pig liver TGase were added. Following incubation for 2 h at 37°C the binding mixtures were centrifuged (12,000 g, 30 seconds at room temperature) and the sedimented cells were washed with RPMI - 1640 medium until no radioactivity was detectable in the supernatant. The washed PBL were transferred before lysis in another polypropylene test tube to avoid the contamination of the final cell lysate with radioactivity not bound to the cells, but adsorbed to the assay test tube walls. PBL lysis was performed by suspending each cellular pellet in 40 µl of distilled water. After addition of 20 µl of a solution containing 15% SDS, 15% β-mercaptoethanol, 190 mM Tris-HCl buffer (pH 6.8) and 30% glycerol, the samples were boiled for two minutes and then analyzed by both TCA-precipitable radioactivity counting and SDS-PAGE (see Fig. 2).

of purified TGase with or without Ca^{2+}. The data shown in Table II clearly indicate that the binding of radioactive SV-IV to human PBL requires the presence of calcium

TABLE III

Effect of the preincubation of $[^{125}I]$ SV-IV with purified guinea pig liver TGase on the $[^{125}I]$ SV-IV binding to human PBL.

Composition of the preincubation mixture	$\begin{bmatrix} 125_I \end{bmatrix}$ SV-IV bound to PBL (cpm $\pm$ S.E.M.)
$[^{125}I]$ SV-IV + TGase	1,400 $\pm$ 120
$[^{125}I]$ SV-IV + TGase + Ca^{2+}	45,220 $\pm$ 3,900

The $[^{125}I]$ SV-IV binding assay was performed as described in Table II after a previous treatment of radioactive SV-IV with TGase in 50 mM Tris-HCl buffer (pH 7.5) containing 5 mM DTT and 120 mM NaCl, in the presence or absence of 2.5 mM CaCl$_2$. At the end of the preincubation (2 h at 37°C) and before PBL addition, 5 mM EGTA was added to the samples in order to inhibit transglutaminase activity during the binding assay.

192

ions. Is the presence of Ca^{2+} sufficient for the binding of native SV-IV to the PBL? Alternatively, is Ca^{2+} a mere cofactor in producing an enzyme-dependent structural modification of SV-IV as a necessary requirement for an effective binding of the protein to the lymphocyte surface? The data shown in Table II strongly suggest that a TGase-dependent modification of SV-IV is indeed demanded to produce a successful binding. This contention is strongly supported by the data, reported in Table III, demonstrating that a marked stimulation of the binding was obtained only when a Ca^{2+} - activated TGase was present during the SV-IV pretreatment. The SDS-PAGE pattern, shown in Fig. 2, indicates that (^{125}I) SV-IV was able to bind to the PBL surface both as high molecular weight form(s) as well as low M.W. form(s). The latter, most probably, are also TGase-modified molecular species. In fact, lane 3 in Fig. 2 clearly shows that a marked binding of low M.W. form(s) of SV-IV can occur also in the absence of Ca^{2+}, but only if the protein is previously treated with active TGase. The binding of SV-IV, occurring in the presence of Ca^{2+} but in the absence of exogenously added TGase, is most probably due to the modifying effect of an endogenous TGase present on the surface of these cells (unpublished data). In conclusion, these data indicate that the low M.W. form(s) of SV-IV, possessing ability to bind to the T cells, are molecular species modified by either exogenous or endogenous Ca^{2+} - activable TGase.

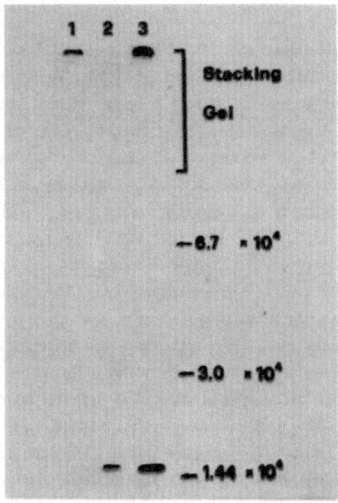

Fig. 2 - Fluorography of the polyacrylamide gel after SDS-electrophoresis, showing the binding to PBL of TGase-modified forms of (^{125}I) SV-IV. Radioactive SV-IV was incubated with PBL and $CaCl_2$ under the binding conditions described in Table II in the presence (lane 1) or absence (lane 2) of purified guinea pig liver TGase. Lane 3 refers to the SV-IV binding assay performed after pretreatment of the radioactive protein with TGase and $CaCl_2$ (see Table III). Equal volumes of samples were loaded onto the gel.

SV-IV inhibits human PBL activation by mitogens.

The presence of SV-IV in PBL culture produces a marked suppression of the activation of T lymphocytes by either various mitogenic factors (ConA, see Table IV; PHA, SEB, OKT3, data not shown) or by other allogenic lymphocytes in mixed cultures (data not shown). Preliminary experiments indicate that also UG displays in vitro powerful immunosuppressive properties. All the results on the immunosuppressive properties of both SV-IV and UG, obtained by using human PBL, were completely reproducible with rat PBL.

Studies are in progress to investigate on the possible involvement of the lymphocyte surface TGase in such immunomodulatory phenomenon.

TABLE IV

Inhibition of human peripheral blood lymphocyte (PBL) response to Con A stimulation by SV-IV.

LYMPHOCYTE TREATMENT	[3H] Thymidine incorporation into lymphocytes (cpm ± S.E.M.)	
	- Con A	+ Con A
PBL 1 (control)	2,380 ± 150	51,642 ± 8,346
PBL 1* + SV-IV	2,072 ± 146	44,600 ± 543
PBL 1 + SV-IV	610 ± 130	393 ± 169
PBL 2 (control)	3,043 ± 616	46,660 ± 2,436
PBL 2* + SV-IV	2,320 ± 508	44,375 ± 730
PBL 2 + SV-IV	621 ± 140	558 ± 102

10^6 cells/ml PBL, obtained from blood of two normal individuals (PBL 1 and 2) by Ficoll-Paque fractionation, were separately treated by incubation for 2 h at 37°C in incomplete RPMI-1640 medium (without glutamine, FCS and antibiotics) in the absence or presence of 80 μg SV-IV. At the end of treatment an aliquot (0.3 ml) of the cell sample containing SV-IV was washed three times with 10 ml of incomplete RPMI-1640 medium (PBL 1* and PBL2*); the cells, after washing, were finally resuspended in 0.3 ml of complete RPMI-1640 medium (containing 2 mM L-glutamine, 10% FCS, 100 I.U./ml penicillin and 100 μg/ml streptomycin). Another 0.3 ml of SV-IV treated cells or 0.3 ml of untreated (control) cells were not washed and appropriate amounts of the components required to complete the RPMI-1640 medium were added. 0.1 ml of each sample (control, SV-IV-treated PBL, SV-IV-treated and then washed PBL (*)) were mixed in the flat bottom wells of a sterile microtiter plate with 0.1 ml of complete RPMI-1640 medium containing when required 25 μg/ml of highly purified Con A, and incubated at 37°C in a CO_2 incubator. After 2.5 days, 1 μCi of (3H) thymidine (specific activity, 2 Ci/mmol, Amersham) was added to each well and the culture continued for another 12 hours. Each culture was done in triplicate. Cell harvesting and washing, filtration and counting of the samples were performed as described in the legend to Fig. 1.

CONCLUSIONS

On the basis of all these data it seems that the main function of the seminal plasma is to ensure, to the cells of vertebrate germ line, a good supply of factors necessary to protect them against a possible immunological aggression in the female genital tract. In this context, we have evidence that SV-IV, TGase and UG can play an important role both in masking the surface immunogenicity of the epididymal spermatozoa and in suppressing the activity of the immunocompetent cells occurring in the reproductive tracts of both sexes.

REFERENCES

1) Hogart P J. Immunological aspects of mammalian reproduction. Blackie, Glasgow and London, 1982.

2) James K and Hargreave T B. Immunosuppression by seminal plasma and its possible clinical significance. Immunology Today 1984: 5; 357-63.

3) Dias J A. Transglutaminase activity in testicular homogenates and Serum-free Sertoli cell culture. Biol Reprod 1985: 33; 835-43.

4) Tung K S K. Autoimmunity of the testis. In: Immunological Aspects of Infertility and Fertility Regulation. Dhindsa D S and Schumacher G F B (eds). North-Holland, 1980: pp. 33-92.

5) Olsen G P and Shields J W. Seminal lymphocytes, plasma and AIDS. Nature 1984: 309; 116-17.

6) Williams-Ashman H G. Transglutaminases and the clotting of mammalian seminal fluids. Mol Cell Biochem 1984: 58; 51-61.

7) Porta R, Esposito C, De Santis A, Fusco A, Iannone M and Metafora S. Sperm maturation in human semen: role of transglutaminase-mediated reactions. Biol Reprod 1986, in press.

8) Mukherjee D G, Agrawal A K, Manjunath R and Mukherjee A B. Suppression of epididymal sperm antigenicity in the rabbit by uteroglobin and transglutaminase in vitro. Science 1983: 219; 989-91.

9) Manjunath R, Chung S I and Mukherjee A B. Crosslinking of uteroglobin by transglutaminase. Biochem Biophys Res Commun 1984: 121; 400-07.

10) Metafora S, Facchiano F, Facchiano A, Esposito C, Peluso G and Porta R. FEBS Lett, 1986, submitted.

11) Ostrowski M C, Kistler M K and Kistler W S. Purification and cell-free synthesis of a major protein from rat seminal vesicle secretion. A potential marker for androgen action. J Biol Chem 1979: 254; 383-90.

12) Pan Yu-Ching E and Li, S S - L. Structure of secretory protein IV from rat seminal vesicles. Int J Pept Prot Res 1982: 20; 177-87.

13) Higgins S J and Burchell J M. Effects of testosterone on messenger ribonucleic acid and protein synthesis in rat seminal vesicle. Biochem J, 1978: 174; 543-51.

14) Abrescia P, Lombardi G, De Rosa M, Quagliozzi L, Guardiola J and Metafora S. Identification and preliminary characterization of a sperm-binding protein in normal human semen. J Reprod Fertil 1985: 73; 71-77.

15) Hunter W M. Radioimmunoassay. In: Handbook of experimental immunology. Weir D M editor, Third edition. Blackwell Scientific Publications 1978; pp. 14.1-14.40.

16) Connellan J M, Chung S I, Whetzel N K, Bradley L M and Folk J E. Structural properties of guinea pig liver transglutaminase. J Biol Chem 1971: 246; 1093-098.

17) Paonessa G, Metafora S, Tajana G, Abrescia P, De Santis A, Gentile V and Porta R. Transglutaminase-mediated modifications of the rat sperm surface in vitro. Science, 1984: 226; 852-55.

18) Thorsby E and Bratlie A. In: Histocompatibility testing. Terasaki I Editor. Munksgaard, Copenhagen, 1970:p. 655.

195

19) Laemmli U K. Cleavage of structural proteins during the assembly of the head of bacteriophage T_4. Nature 1970: 227; 680-85.

20) O'Farrell P H. High resolution two-dimension electrophoresis of protein. J Biol Chem 1975: 250; 4007-021.

21) Bonner W M and Laskey R A. A film detection method for tritium-labelled proteins and nucleic acids in polyacrylamide gels. Eur J Biochem 1974: 46; 83-88.

22) Lowry O H, Rosenbrough N J, Farr A L and Randall R J. Protein measurement with the Folin phenol reagent. J Biol Chem 1951: 193; 265-75.

Hydrolytic enzyme detected in the fluid and in spermatozoa from different portions of human epididymis

J.F. GUERIN, J.C. CZYBA

Laboratoire de Biologie de la Reproduction et du Développement, Faculté de Médecine, Lyon, France

INTRODUCTION

Various glycosidases have been described in epididymal secretions from different species (1-4). However the situation is not well known in the human male, because of evident difficulties to obtain epididymal samples from healthy and fertile subjects.In a previous study (5), we compared numerous enzyme activities in semen from azoospermic patients with testicular dysfunction on one hand, and from vasectomized men, on the other hand. Only one enzyme - α glucosidase - was missing in that last group. These data indicated that most of the enzymes present in epididymal fluid are also secreted by sexual accessory glands, and are thereby difficult to isolate.

In the present work, we have used a microtechnique to determine the hydrolytic activities present in the successive portions of human genital ways. α glucosidase activity was measured more precisely because of its interest as a new epididymal marker.

MATERIAL & METHODS

Epididymal fluid was obtained from 25 patients where a surgical scrotal investigation was performed because of a suspected obstruction of genital tractus : all of them were azoospermic and had normal levels of plasma FSH.The catheterizations were made upstream of the obstructive zone each time this one was macroscopically detectable. When they were found, spermatozoa were separated from epididymal fluid by gentle centrifugation (400 g, 10 min) and resuspended in a hypotonic salt medium.

In some cases, we could obtain some portions of epididymides (18 from 8 subjects) containing no spermatozoa, which were homogenised in a buffered salt solution. We also obtained two healthy epididymides from a young man who died suddenly. After high speed centrifugation, the supernatants were submitted to the same enzymatic analysis as epididymal fluid and sperm suspensions.

Secretions of sexual glands (prostate and seminal vesicles) were investigated by the bias of semen obtained from 25 vasectomized men.

Enzymatic analysis : The API ZYM microtechnique was used as previously described (6) to estimate simultaneously 19 hydrolase activities (mostly glycosidases and phosphatases). This colourimetric semi-quantitative technique was applied on 11 epididymal sperm suspensions, 17 epididymal fluids, and 18 homogenates.

α glucosidase activity was measured in some homogenates and in

semen from vasectomized men by a spectrophometric technique based
on the release of p-nitrophenol from a synthetic susbtrate (7). We
have measured neutral and acid forms of the enzyme by using selective inhibitors (8) : Sodium Dodecyl Sulfate (SDS) inhibits the acid
isoenzyme whereas Maltotriose (MTT) inhibits the neutral isoenzyme.

RESULTS

The principal data are summarized in Table I : in the fluid from
rete testis and efferent ducts we detected weak activities of este-
rase, leucine arylamidase, phospho amidase and high activities of
acid phosphatase. No marked changes occurred in the initial portion of epididymal duct. By contrast, in the proximal corpus the
enzyme activities present in rete testis fluid were dramatically increased, whereas other enzymes were detected, namely β galactosidase,
N acetyl β D glucosaminidase, β glucuronidase, and α glucosidase. In
the following portions of the genital tractus, i.e. epididymal cauda
and vas deferens, the enzymatic profile underwent only minor variations.

Secretions of accessory glands contained the same enzyme activities, with the exception of valine arylamidase (which appeared) and
α glucosidase (which disappeared).

We noted a perfect similarity between enzymatic profiles of epididymal fluid and spermatozoa, for the same anatomical level.

In homogenates we also revealed the same enzyme activities as
in the epididymal fluids at the corresponding levels, with one exception : β D galactosidase was much more active than in the fluid.

The values of spectrophotometric measurements for α glucosidase
are presented in Table II. For homogenates, we have only considered
the two samples devoid of scrotal pathology, i.e. from young man who
suddenly died. This quantitative analysis confirms the weak activity
of α glucosidase present in the epididymal caput, and its dramatic
increase on the corpus. Then the enzyme activity decreased in the
following portions (cauda and vas deferens) but remained measurable.

By using selective inhibitors we showed that the "total" activity

Table I . Hydrolases detected in fluid from human genital ways
           (17 samples)

ENZYMES	rete testis efferent ducts	Epididymis			vas deferens	accessory glands
		caput	corpus	cauda		
esterase	+	+	+	+	+	++
leucine arylamidase	+	+	+	+	+	++
phosphatase acide	+	+	++	++	++	++
phospho amidase	+	+	++	++	++	++
esterase lipase	?	?	+	+	+	+
β galactosidase	0	?	+	+	+	?
N acetyl βD glucosaminidase	0	?	++	++	++	++
valine arylamidase	0	?	?	?	?	++
phosphatase alcaline	0	0	+	+	+	++
β glucoronidase	0	0	++	++	++	++
α glucosidase	0	0	++	++	++	0
α fucosidase	0	0	?	?	?	+

of α glucosidase in the epididymis represented mainly the neutral iso form. In contrast, the total activity detected in semen from vasectomized men represented predominantly the acid form of α glucosidase, whereas the neutral activity was very low.

Table II. Activities of α glucosidase isoforms in homogenates from epididymis and vas deferens, compared to semen from vasectomized men.

	EPIDIDYMIS (a)			VAS (a) DEFERENS	SEMEN FROM VASECTOMIZED MEN (b)
	CAPUT	CORPUS	CAUDE		
TOTAL ACTIVITY	5.9	36	10	8	9.8
NEUTRAL ISOFORM	1.8	34	12	8	5.2
ACID ISOFORM	—	—	—	—	19

(a): values expressed in mI.U./g of wet organ
(mean of 2 observations)
(b): values expressed in mI.U./ml X volume of ejaculate
(mean of 5 observations).

DISCUSSION

Enzymes found in epididymal plasma may have different origins : (1) they originate from upper genital ways : seminiferous tubules or rete testis ; (2) they are synthetised by the epididymis epithelium itself ; (3) they are released from damaged spermatozoa. Our study does not permit to answer this question precisely. We realise that it is limited because we only investigated one class of enzymes (hydrolases), and not metabolic enzymes as dehydrogenases and kinases, which originate mostly from spermatozoa.

Our results reveal a poor content of enzymes in fluid from rete testis and epididymal caput, contrasting with the apparition or increase of numerous enzyme activities in the initial portion of epididymal corpus. An important resorbtion of fluid occurs in the caput of epididymis, which could explain the relative increase of enzyme activities. Nevertheless, it is interesting to remark that in this region important events occur concerning maturation of spermatozoa : the cells become motile and acquire their fertilizing capacity (9). Glycosidases may play an important role in relation with changes in the composition of sperm membranes. In contrast, there was no further important variations concerning enzymatic composition in epididymal cauda and vas deferens.The similarity between enzymatic activities of epididymal fluid and homogenates whether sperm are present or not in the lumen, confirms thast most of hydrolase originate from epididymal secretions and not from spermatozoa (3,10). We also observed a remarkable similarity between enzyme content of epididymal fluid and that of spermatozoa. These data are in agreement and confirm our earlier study (6), which indicated that in ejaculated semen most of enzymes originated from seminal plasma and were adsorbed on sperm membranes, the reciprocal event being less important.

A release of enzymes from damaged spermatozoa is not excluded and occurs likely since a variable proportion of gametes die and are destroyed in the epididymal lumen. But the activity of these enzymes is probably much less important than that of enzymes secreted by epididymal cells.

Analysis of semen from vasectomized men traducts perfectly the enzyme content of sexual accessory glands secretions (prostate and seminal vesicles together), without any participation of those from genital ways. The comparison with epididymal fluid shows that accessory glands secrete the same hydrolases as epididymis, with higher activities. Fortunately one enzyme - α glucosidase - is secreted specifically by epididymis and constitutes thereby a reliable marker of epididymal function (11).

In fact, a small residual activity remains in semen after vasectomy, which is mainly constituted of acid isoform confirming other studies (12). In contrast the epididymal secretions contain mainly neutral α glucosidase, which represents thereby a good selective marker of epididymal secretions. We have also precised the site of secretion : the activity appears in the proximal corpus, but does not seem to be strictly limited to that protion, since we also detected a reduced but undoubtful enzyme secretion in the epididymal cauda.

REFERENCES

1. Conchie J, Mann T. Glycosidases in mammalian sperm and seminal plasma. Nature 1957: 1190-93.
2. Snaith SM, Levvy G. Purification and properties of α D-mannosidase from rat epididymis. Biochem J 1969: 25-32.
3. Moneim KA, Glover TD. Comparative histochemical localization of lysosomal enzymes in mammalian epididymides. J Anat 1972a: 111; 437-52.
4. Mann T, Lutwak-Mann C. Male reproductive function and semen. Springer-Verlag Berlin Heidelberg. New-York, 1981.
5. Guerin JF, Rollet J, Perrin P, Menezo Y, Orgiazzi A, Czyba JC. Enzymes in the seminal plasma from azoospermic men : correlation with the origin of their azoospermia. Fertil Steril 1981b: 36; 368-72.
6. Guerin JF, MENEZO Y, Czyba JC. Enzyme comparative study of spermatozoa and seminal plasma in normal and subfertile man. Arch Androl 1979: 3; 251-57.
7. Chapdelaine P, Tremblay RR, Dube JY. P-nitrophenyl-α-D- pyranoside as substrate for measurement of maltase activity in human semen. Clin Chem 1978: 24; 208-11.
8. Paquin R, Chapdelaine P, Dube JY, Tremblay RR. Similar biochemical properties of human seminal plasma and epididymal α 1,4-glucosidase. J Androl 1984:5; 277-82.
9. Dacheux JL, Paquignon M. Relations between the fertilizing ability, motility and metabolism of human spermatozoa. Reprod Nutr Develop 1980: 20; 1085-99.
10. Jones R, Glover TD. The collection and composition of epididymal plasma from the cauda epididymis of the rat. J Reprod Fertil 1973a: 34; 395-403.
11. Guerin JF, Ben Ali H, Rollet J, Souchier C, Czyba JC. α -glucosidase as a specific epididymal enzyme marker:its validity for the etiologic diagnosis of azoospermia.
12. Tremblay RR, Chapdelaine P, Dube JY. Neutral α 1,4-glucosidase in human seminal plasma : molecular forms in varicocele and after vasectomy. Fertil Steril 1982: 38; 344-48.

Morphological study on the interaction between enterococci and human spermatozoa

L. ACCINNI, R. GIACOMELLI[1], A. RADICIONI[2], G. TONIETTI[1], F. DONDERO[2]

Institute of Experimental Medicine, National Research Council (CNR), Rome, Italy; [1]Institute of Clinical Immunology, University of L'Aquila, L'Aquila, Italy; [2]Laboratory of Seminology and Immunology of Reproduction, V Clinica Medica, University of Rome "La Sapienza", Rome, Italy

§ This work was in part supported by CNR grant N. 85.02395.44

INTRODUCTION

It has become increasingly apparent that sperm infec= tions may play a role in the genesis of male fertility impairment or infertility (1). Although numerous clinical and laboratory data suggest a relationship between these two pathological conditions, conflicting results exist con= cerning the real incidence and the mechanisms by which microorganisms can alter the reproductive function. Besides the presence of microorganisms, other changes of the semi= nal parameters are frequently observed in the course of these infections, such as alteration of fluidity of the seminal plasma, sperm hypomotility and/or diskinesia, tera= tozoospermia, increased leukospermia, and sperm agglutina= tion areas. The ability of microorganisms to adhere to human spermatozoa has been demonstrated in 'in vitro' ex= periments (2,3,4) and shape and motility alterations of the spermatozoa have also been reported.

To further clarify the relation of sperm infections with male infertility, we examined by light and electron microscopy the effects, so far unexplored, of the incuba= tion of Enterococci, which among gram positive bacteria are the most frequently involved in this pathology, with human spermatozoa.

MATERIALS AND METHODS

Enterococci were isolated from infected seminal fluids

201

and identified by specific stains and selective cultivation
methods by Dr. M.Sappa, Microbiology Division, Central Lab=
oratory, Policlinico Umberto I, Rome. Bacteria were washed
three times in Earle's solution and then suspended in the
same medium. The concentration of Enterococci per ml, de=
termined immediately before the adherence test, was adjusted
to obtain suspensions containing 10^4 to 10^7 bacteria per ml.

Standard populations of motile <u>spermatozoa</u> were iso=
lated from sterile semen samples by the 'swim up' technique
and divided into several aliquots containing 50×10^6 sperm/ml,
some of which were used for the incubation experiments while
others were kept as controls without the addition of bacteria.

For the 'adherence test', samples (1 ml) of sperm were
mixed with equal volumes of increasing concentrations of
Enterococci and incubated in a water bath shaker at 37°C for
12 hr. At sequential time intervals (30 min, 1 hr, 2 hr,
and 12 hr) fresh bacteria-sperm suspensions were examined
with a phase contrast microscope, while aliquots were proc=
essed for transmission electron microscopy. For the latter,
the suspensions were fixed in picric acid-formaldehyde solu=
tion (5) for 2 hr, postfixed in 1% unbuffered OsO_4, dehy=
drated in graded ethanol, embedded in Epon 812, and exam=
ined with a Siemens 1A electron microscope.

RESULTS

Both phase contrast and electron microscopic observa=
tions showed frequent adherence between bacteria and sper=
matozoa (figs.1,2,and 3). No preferential site of adhesion
was observed. In fact, bacteria were found attached to all
segments of the spermatozoon. (fig.2).

The time course of the adherence phenomenon could be
divided into three phases. During the first one, lasting
about 1 hr, the attachment of bacteria appeared unstable
and the movements of the spermatozoa seemed altered. In the
second phase, the adhesion of bacteria appeared firmer and
the movements of the spermatozoa slower, while sperm agglu=
tination initiated. The third phase, which in part over=
lapped phase 2, was characterized by a progressive increase
in the number of spermatozoa carrying bacteria and in the
number of adhered bacteria per spermatozoon.

In the controls, agglutination areas were never ob=
served and sperm motility did not decrease within the first
two hours, while a physiological reduction initiated after=
wards.

No relevant variations in the physicochemical charac=

202

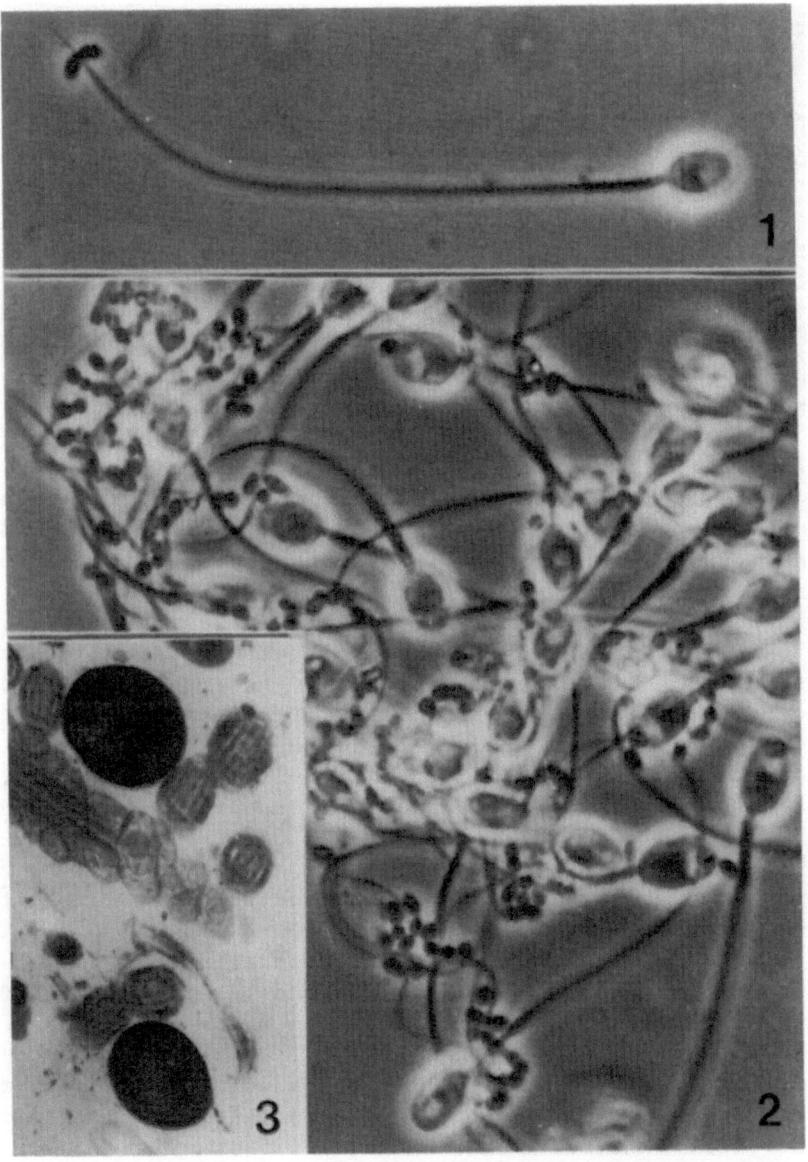

Figs.1 and 2: phase contrast microscopy. Fig.1 shows a
spermatozoon with two bacteria attached to the end of the
tail, while in fig.2 a large agglutination area is visible.
Note the various sites of bacteria adhesion to the spermat=
ozoa; x 3,000. Fig.3: the electron micrograph illustrates
at high magnification two bacteria attached to several
spermatozoan tails; x 35,000.

teristics of the suspension medium were detected at all time intervals.

DISCUSSION AND CONCLUSIONS

Our study clearly indicates that Enterococci adhere to human spermatozoa 'in vitro' under appropriate conditions. The degree of adherence and its effects on sperm motility seem related to the concentration of bacteria and incubation time.

The mechanisms of Enterococci-sperm adhesion are still under study. It may be postulated that they involve the presence of receptors and/or cell surface electric charges (6); however, several other factors may be of relevance. The most interesting finding of our study is the alteration of sperm motility, followed by sperm agglutination, which seems strictly correlated with bacteria adhesion. In fact, it is evident that such changes may play an important role for the induction of infertility frequently associated with sperm infections (1).

In conclusion, the results of this preliminary study seem to indicate that if one extrapolates 'in vitro' exper= iments into an 'in vivo' situation (7), sperm infections by Enterococci may provoke male infertility through the alter= ation of sperm motility and formation of sperm agglutination areas, caused by the adherence of bacteria to spermatozoa. The factors sustaining this adherence are presently under study.

REFERENCES

1. Buvat J,Buvat-Herbatt M. Infection du sperm et fertilité. Contraception,Fertil.,Sexual. 1985:12;401-409.
2. Friberg J,Fullan N. Attachment of Escherichia Coli to human spermatozoa. Am.J.Obstet.Gynecol. 1983:146;465-467.
3. Gnarpe H,Friberg J. T Mycoplasmas on spermatozoa and infertility. Nature 1973:245;97-98.
4. James AN,Knox JM,Williams RP. Attachment of Gonococci to sperm.Influence of physical and chemical factors. Brit.J. Vener.Dis. 1976:52;128-135.
5. Stefanini M,DeMartino C,Zamboni L. Fixation of ejaculated spermatozoa for electron microscopy. Nature 1967:216;173-175.
6. Savage DC,Fletcher M (Eds). Bacterial adhesion:mechanisms and physiological significance. New York,Plenum Press,1985.
7. Radicioni A,Giacomelli R,Lenzi A,Vercelloni B,Accinni L, Dondero F. Il test di aderenza batterica nello studio della infertilità maschile da infezioni del tratto geni= tale. Proceed.XIII Nat.Congr.S.I.F.E.S.,Salsomaggiore, May 22-24,1986.

A scanning electron microscopic study on the selection of spermatozoa for artificial insemination

A. CAGGIATI, A. LENZI[1], F. CLARONI[1], L. GANDINI[1], A. LUZI, F. DONDERO[1], P.M. MOTTA

Department of Anatomy; [1]Laboratory of Immunology of Reproduction and Seminology, 5th Medical Clinic, University of Rome "La Sapienza", Rome, Italy

INTRODUCTION

Semen manipulation for artificial insemination purposes (AIH, AID, FIV-ET, GIFT) represents one of the expressions of the high-level of laboratory technology in andrology. These techniques permit the clinician to tackle and resolve cases resistent to systemic therapy.

Scanning electron microscopy (SEM) allows an ultrastructural threedimensional view of the biological specimen and in seminology the SEM demonstrated usefulness to study spermatozoa by hundreds on a single sample (1).

The aim of the present study was to verify with the SEM the eventual damage induced by the above—mentioned methods of semen preparation, overcoming limits of light microscopy (2).

MATERIAL and METHODS

Many techniques for selection of spermatozoa populations are routinely used in our Laboratory (3). These techniques aim at obtaining a pool of normal motile spermatozoa, with good concentration in a little quantity of culture buffer. For the purposes of this work, three of these techniques were chosen on the basis of the presence in their methodological steps of only one kind of possible cell trauma.
A) Filtration in glass wool (Fig. 1): semen is filtered through columns filled with glass wool and collected in aliquots. The possible cell trauma could be induced by the contact with the glass fibres.
B) Swim—up migration (Fig. 2): the buffer is

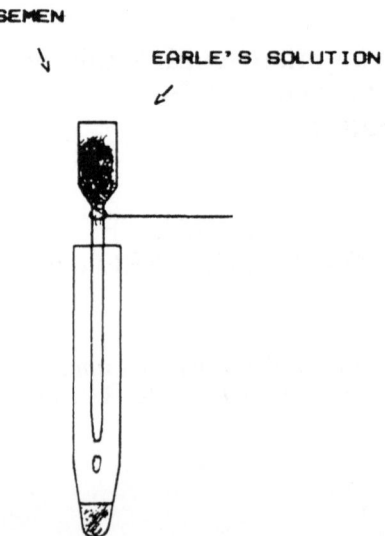

FIGURE 1. Filtration in glass wool

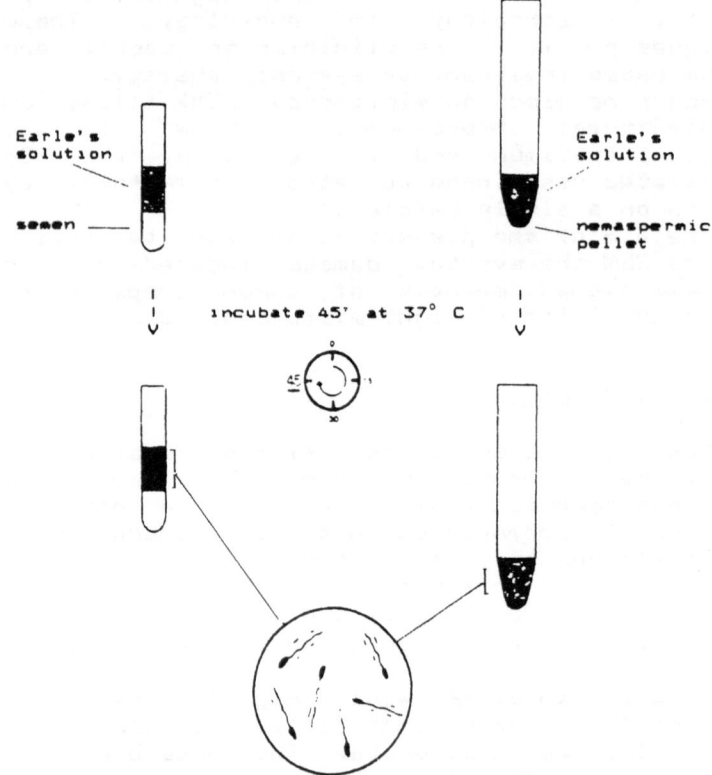

FIGURE 2. Swim-up migration and pellet swim-up

layered on aliquots of semen and motile spermatozoa migrate actively into the buffer. The possible cell trauma could be induced during the migration between the two media with different density.

C) Pellet swim-up (Fig. 2): the buffer is layered on a pellet obtained after centrifugation of the semen; then motile spermatozoa migrate actively in the buffer. The possible cell trauma could be induced by the centrifugation.

The above-mentioned techniques were carried out on seminal samples of healthy fertile subjects to exclude possibilities of interferences deriving from high percentage of teratozoospermia.

After seminological manipulation, semen specimens were fixed with 2.5% glutaraldehyde gently added to the buffer medium. Fixed spermatozoa were then dropped on a cover glass previously treated with polylisine. Cover glasses were then dehydrated in ethanol, critically point dried and sputtered with gold. The observations were performed with the Scanning Electron Microscope Cambridge Stereoscan 150, operating at 20 Kv.

RESULTS

No statistical analysis was possible on this preliminary series of samples, but the following considerations could be underlined.

Control specimens showed normal sperm morphology (Fig. 3-4). Spermatozoa selected by the swim-up migration technique showed a well conserved morphological feature. Cytoplasmatic blebs and plasmalemma defects were rarely observed (Fig. 5-6).

Spermatozoa selected with the pellet swim-up technique showed more frequent and serious membrane defects (Fig. 7). Furthermore, tail damaging were frequently observed (Fig. 8).

Spermatozoa filtered across glass wool showed frequently head fragmentation and tail section. Moreover, narrowed fragments of glass wool were present on the sperm surface (Fig. 9-10).

DISCUSSION and CONCLUSION

On the basis of these preliminary findings the glass wool filtration technique for spermatozoa selection seems to be the more traumatic one. On the contrary, the swim-up migration induces a lower number of damaged spermatozoa. However, in

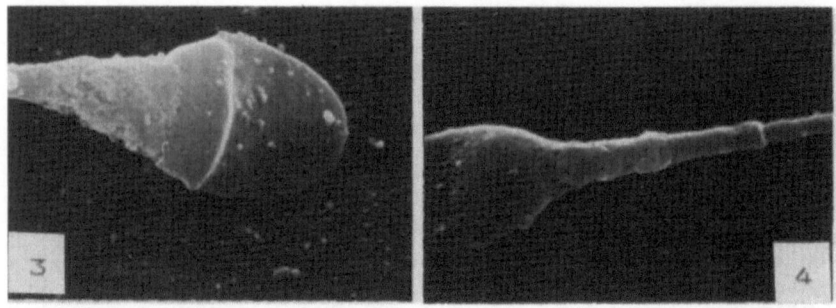

FIGURE 3. Normal spermatozoa from control specimens
FIGURE 4. Normal spermatozoa from control specimens

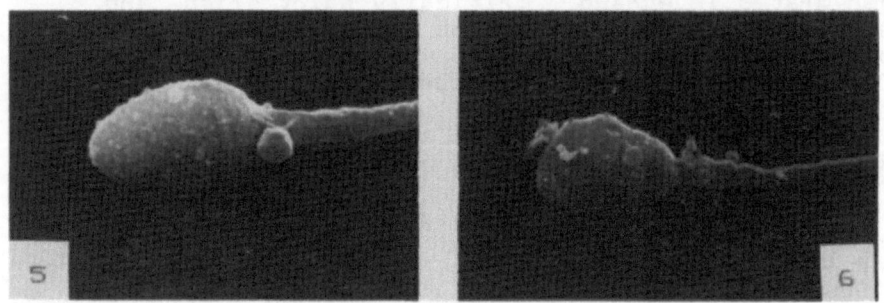

FIGURE 5. Cytoplasmatic blebs in spermatozoa selec-
          ted by the swim-up migration technique
FIGURE 6. Plasmalemma defects in spermatozoa selec-
          ted by the swim-up migration technique

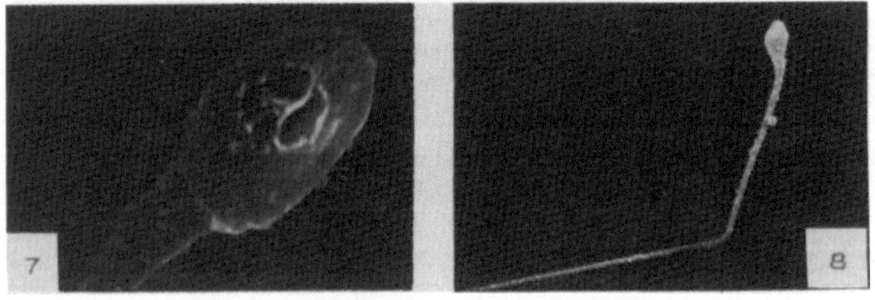

FIGURE 7. Membrane defects in spermatozoa selec-
          ted by pellet swim-up technique
FIGURE 8. Tail damaging in spermatozoa selected
          by pellet swim-up technique

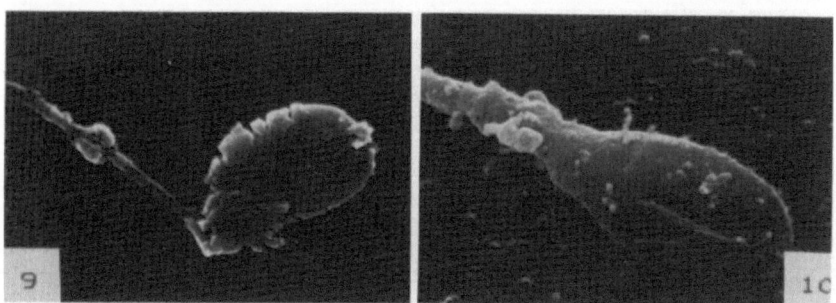

FIGURE 9. Head fragmentation in spermatozoa filtered
           across glass wool
FIGURE 10. Fragments of glass wool on the surface of
           spermatozoa filtered across glass wool

our opinion, wider casistics should be studied in
order to obtain a better discrimination between
morphological abnormalities due to the seminological manipulation procedure and those eventually
caused by the electron microscope preparation.

REFERENCES

1) Gould KG, Martin DE, Hafez ESE. Mammalian spermatozoa. In: Scanning Elecron Microscopy Atlas
   of   Mammalian Reproduction,   ed. Hafez ESE,
   Tokio, Igaku Shoiu Ltd, 1975: 42-57.
2) Sherman JK, Dmowsky WP. Effect of isolation
   by albumin density gradients on ultrastructure
   of human spermatozoa.   Fertil Steril 1982: 38;
   460-64.
3) Dondero F, Lenzi A, Gandini L, Claroni F, Lom-
   bardo F, Benagiano G. Clinical application of
   techniques for selection of spermatozoa in
   homologous artificial insemination. New Trends
   Gynaecol Obstet 1986: 2; 199-210.

209

Localization of antisperm antibodies on sperm surface by indirect and direct immunofluorescence test using sperm suspensions

R. ROMANO, F. FRANCAVILLA, R. SANTUCCI, A. BARONE, P. CATIGNANI, G. PROPERZI, A. FABBRINI

Department of Internal Medicine, University of L'Aquila, L'Aquila, Italy

INTRODUCTION

The indirect immunofluorescence (IIF) test for the detection of antisperm-antibodies on washed and methanol-fixed sperm smears (1) has not gained wide acceptance because of two major reasons. Firstly, it detects antibodies against internal sperm antigens, which have not a role in infertility (2); secondly, the great amount of non-specific background represents a major obstacle for the reproducibility and realibility of the test (3).

We have carried out an IIF test using sperm suspensions instead of fixed sperm smears, in order to obtain a reproducible and sensitive test system valuable for the detection of circulating sperm-surface related antibodies. A direct immunofluorescence (DIF) test on sperm suspensions is also proposed.

MATERIALS AND METHODS

IIF was carried out on 20 male sera with previously demonstrated sperm-agglutinating (SA) activity (titres: 1:16 - 1:2048) by the modified slide agglutination test (MSAT) (4), and on 25 negative controls.

DIF test was carried out on sperm suspensions from the same patients with serum SA activity (14 cases) and on 15 controls.

Procedure of IIF test. Motile sperm suspensions were obtained from donor semen samples by swim-up procedure after dilution with BWW and centrifugation. Sperm suspensions were divided into aliquots containing about 20×10^6 spermatozoa and each pellet, obtained by centrifugation, was resuspended and incubated with an undiluted serum sample for 60 min. at room temperature. After 2 washings in PBS, each pellet was resuspended and divided into 3 aliquots. The pellets, obtained by centrifugation were resuspended and incubated with FITC labeled $F(ab')_2$ antisera against human IgG, -IgA and -IgM (Kallestad). The proper dilution for FITC labeled antisera resulted 1:10 for anti

IgG and anti IgA and 1:5 for anti IgM. After 3 washings, supernatants were discarded leaving only a drop to allow resuspension. Results were scored on a Zeiss epifluorescence microscope. Fluorescence was graded + to +++.

Procedure of DIF test. Semen specimens from patients (or controls) were washed 3 times in PBS by centrifugation and each sperm suspension was divided into 3 aliquots (each containing about 5×10^6 spermatozoa). The pellets obtained by centrifugation, were resuspended and incubated with FITC labeled $F(ab')_2$ antisera against human IgG, -IgA and IgM (Kallestad). For the subsequent steps see the procedure of IIF test.

RESULTS

IIF test gave positive results in all sera with SA activity and negative results in all sera without SA activity. In relation to "tail-tail" or "mixed" SA activity (19 cases), IF-reactivity for IgG was found and the fluorescent stain constantly appeared in a granular pattern along the sperm tail and most often on the head surface too (Fig.1). In relation to high titre of "head-head" SA activity (1 case), IgM were involved in IF-reactivity and the fluorescent stain appeared localized on acrosomal surface (Fig.1). Positive reactions constantly involved the total number of spermatozoa or in cases of weaker reactant sera (+) a major percentage of them.

DIF test gave positive results in 13 of the 14 patients with serum SA activity, and negative results in all specimens from controls. Both IgG and IgA resulted involved in IF-reactivity in 53.8% of positive cases, only IgG in 30.8% and only IgA in 15.4%. Fluorescent stain constantly appeared in a granular pattern along the sperm tail and often on the sperm head surface too, both with IgG and IgA (Fig.1).

DISCUSSION

IIF test on sperm suspensions specifically and sensitively detects sperm-surface related antibodies, as demonstrated by the comparison of its results with the occurrence of serum SA activity. Moreover, the granular fluorescent stain, which lines the sperm-surface, is stronghly suggestive of surface related fluorescence and is different from all the patterns of positivity previously described using methanol-fixed sperm smears (1,2,3). The use of cell suspensions appears to increase the sensitivity of IIF test for the detection of cell membrane related Ig, since positive results had never been reported on unfixed sperm smears (1). Finally, the relation found among class of Ig, fluorescent stain and type of SA activity further confirms that IF reactivity and SA activity are related to the same kind of antibodies. The relation found between class of Ig at IIF test and type of SA activity is in agreement with a previous

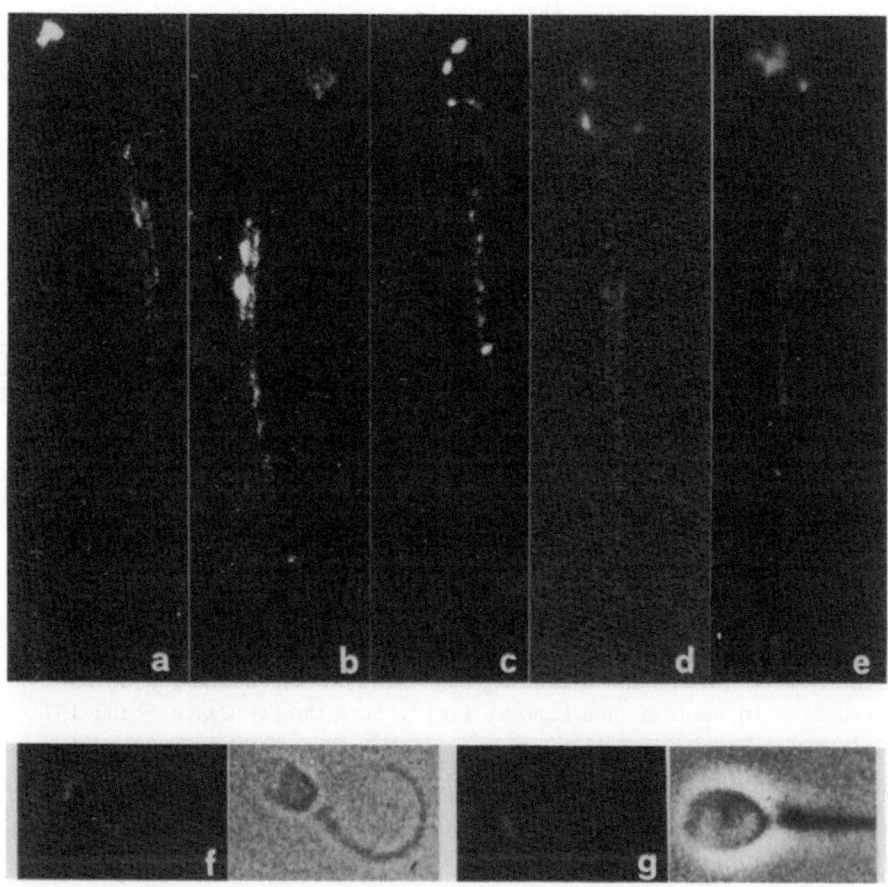

Fig.1. Positive reactions with antisera against IgG at IIF (a,b,c,d)
and at DIF (e), and positive reaction with antisera against IgM
at IIF (f,g).

report by Friberg, who demonstrated the "head-tail" and the
"head-head" type of SA activity respectively in the IgG and IgM
fractions in male sera (5).

Positive results with DIF test for the detection of
antisperm-antibodies are firstly described here. DIF on
sperm-suspensions appears both sensitive and specific for the
detection of sperm membrane bound antibodies, when its results are
compared with the occurrence of serum SA activity. Antisperm
antibodies play a role in infertility only when they occur in the

reproductive tract, where they are largely absorbed onto the sperm surface. Therefore, a great interest is roused by techniques proposed for the detection of sperm-associated antibodies. With this respect, DIF may have a useful role in the clinical evaluation of immunological infertility.

REFERENCES

Hjort T and Hansen B. Immunofluorescent studies on human spermatozoa. Clin Exp Immunol 1971: 8; 9-23

Tung KSK. Human sperm antigens and antisperm antibodies. Clin Exp Immunol 1975: 20; 93-99

Boettcher B, Hjort T, Rumke P, Shulman S and Vyazov O. Auto- and iso-antibodies to antigens of the human reproductive system. Results of an international comparative study. Clin Exp Immunol 1977: 30; 173-180

Francavilla F, Catignani P, Romano R, Santucci R, Francavilla S and Santiemma V. Modification of the slide agglutination test for the detection of sperm-agglutinins. Andrologia 1983: 15; 699-704

Friberg J. Clinical and immunological studies on sperm-agglutinating antibodies in seminal and seminal fluid. Acta Obstet Gynec Scand 1974: 36 (suppl); 21-76

Angiotensin converting enzyme release by human spermatozoa during capacitation

C. FORESTA, M. INDINO, G. SCANELLI, A. CARET-TO, C. SCANDELLARI

Istituto di Semeiotica Medica, Università di Padova, Padova, Italy

INTRODUCTION

It has been observed that human seminal plasma shows a considerably high angiotensin-converting activity (1,2). The source and the localization of Angiotensin-Converting-Enzyme (ACE) in the human reproductive tract have not been investigated as yet, but experimental studies, performed in animals, strongly suggest that this enzyme in synthesized within the reproductive organs, including testis and epididymis (3-6). Testicular ACE appears in the developing testis during spermatogenic maturation, of which it is considered a marker (4), and autoradiographic studies demonstrate that this enzyme is associated with spermatids, residual bodies, epididymal spermatozoa and cytoplasmic droplets (3-6), while non-germinal testicular cells exhibit a very low enzyme activity (3). Testicular and epididymal ACE do exist in a particulate form (5): particulate ACE has been supposed to change into its soluble form when spermatids undergo maturation, during the passage from the testis through the epididymis (5).

It is as yet unknown whether ejaculated human spermatozoa contain ACE and whether the release of this enzyme has a physiological significance.

In this paper we estimated the ACE content of human spermatozoa and studied whether the release of this enzyme may be related to capacitation or acrosome reaction.

MATERIALS AND METHODS

Semen samples obtained from fertile healthy volunteers (aged 22 to 27 years) after three days' sexual abstinence, were allowed to liquify at room temeprature. The following parameters were always analyzed:semen volume and pH, sperm concentration, motility, morphology and viability, measured by eosin red exclusion test. Only specimens with a viability higher than 80% and a motility higher than 60% were used.

The seminal samples were pooled and combined with an equal volume of modified Krebs-Ringer Solution (KRS) containing: 154 mM NaCl, 1 mM

CaCl$_2$, 1 mM KH$_2$PO$_4$, 1 mM MgSO$_4$, 5 mM NaHCO$_3$, 5 mM Glucose and buffered with 20 mM N-2-HydroxyEthylPiperazine-N'-2-EthaneSulphonic acid (HEPES), pH 7.38, at 37°C.

Spermatozoa were washed three times and finally supernatants were discarded and the pellets equally distributed into tubes.

In the first set of tubes, the volume was made up to 2 ml using KRS; to these samples the following substances were added:
a) none;
b) 0.2% Digitonin;
c) 0.1% Triton X-100.

All samples were incubated at 37°C in an atmosphere of air/ 5% CO$_2$. After the incubation time, supernatants were collected by centrifugation and stored at -20°C until ACE was assayed. From samples a) supernatants were collected at 1 hour intervals for six hours; after this incubation, aliquots of the samples were collected and further incubated for 10 minutes with 0.2% Digitonin. From samples b) and c), supernatants were collected after 10 minutes from the addition of Digitonin or Triton X-100, respectively.

In the second set of tubes, the volume was made up to 2 ml using capacitating medium, BWW, containing: 95 mM NaCl, 4.8 mM KCl, 1.7 mM KH$_2$PO$_4$, 2.4 mM MgSO$_4$, 25 mM NaHCO$_3$, 20 mM HEPES, 0.25 mM Na-Piruvate, 21 mM Na-Lactate, 5.6 mM Glucose and added with 3% Bovine Serum Albumin Fraction V.

Samples were incubated for six hours and aliquots of supernatants were collected at 1 hour intervals for six hours.

To evaluate spermatozoa ACE content during BWW-incubation, aliquots of the cellular suspension were added, after six hours of incubation in capacitating conditions, with 0.2% Digitonin and supernatants were collected after 10 minutes of incubation.

ACE was measured spectrophotometrically, according to a modified method of Cushman and Cheung (7), using the synthetic substrate Hyppuril-Hystidil-Leucine, which is not hydrolyzed by most of other tissues peptidases. The incubation mixture consisted of 0.7 ml of 0.1 M borate/sodium/carbonate buffer (pH 7.8), 0.2 ml of 10 mM Hyp-Hys-Leu in the same buffer and 0.1 ml of the sample solution. The final concentration of NaCl in the incubation mixture was 0.8 M. The mixture was incubated at 37°C for 30 minutes. One unit of the enzyme activity was defined as the amount of the enzyme which hydrolyzed 1 uM of Hyp-Hys-leu per minute.

Acrosome reaction was evaluated by means of an indirect immunofluorescent technique, using a monoclonal antibody against a sperm cytokeratin. The method for obtaining the antibody is described in the paper of Wolf et al. (8). Small aliquots of each sample were spread on precleaned glass slides and allowed to dry for a few minutes before they were stored at -80°C until acrosome reaction was evaluated. The presence or absence of sperm acrosome was established by adding to each slide the monoclonal antibody T6 (diluted 1:30); the reaction was revealed by adding subsequently biotin and avidin (diluted 1:40).

Slides were finally mounted in buffered glycerol and examined by a
Leitz Orthoplan fluorescence microscope. Cells were scored positive or ne
gative, depending on the presence of uniform fluorescence restricted to
the acrosomal cap region (acrosome intact) or the absence of the specific
fluorescence (acrosome reacted), respectively.

Results are the mean of five experiments. Data are expressed as the
mean $\pm$ SE. Statistical analyses were performed by using Student's t test
and one way analysis of variance test.

The monoclonal antibody T6 was kindly donated by D.P. Wolf.

RESULTS

Table I shows the effects of 0.2% Digitonin and 0.1% Triton X-100 on
ACE release by human spermatozoa and the percentage of spermatozoa with
acrosome loss. Both detergents determine a similar important release of
ACE by spermatozoa and 100% of acrosome loss.

Fig. 1 shows the ACE release by spermatozoa incubated for 6 hours ei
ther in KRS or in BWW-solution (added with 3% BSA). Spermatozoa incubated
in capacitating conditions release ACE in a time-dependent manner, showing a maximum at the 5th and 6th hour of incubation. Only small amounts
of ACE are released by spermatozoa incubated for the same time in KRS sol
ution.

Fig. 2 shows the effects of 0.2% Digitonin on ACE release by spermatozoa after 6 hours' incubation either in KRS or in 3% BSA-added BWW.
KRS-incubated spermatozoa expressed, after Digitonin treatment, an angiotensin-converting activity comparable to that observed in spermatozoa
after 6 hours' incubation in BWW. Digitonin was not able to elicit any
further release of ACE by spermatozoa incubated for 6 hours in BWW.

Fig. 3 shows the immunofluorescence picture obtained by treating
spermatozoa with the monoclonal antibody T6. The percentage of acrosome
reactions in BWW-incubated spermatozoa was 29 ± 2.1, after 6 hours' incub
ation, while only 7.1 ± 0.8% of KRS-incubated spermatozoa was acrosome
reacted after the same time.

Table I

ACE Activity of Extracts Obtained from Ejaculated Human Spermatozoa

EXTRACTION TECHNIQUE	ACE ACTIVITY U/l	ACROSOME REACTION
None	5.8 + 0.8	3%
Triton X-100	100 + 2.5	100%
Digitonin	107 + 2.3	100%

Effects of Digitonin and Triton X-100 on ACE release and acrosome reaction in human spermatozoa. Each experiment was performed using samples of
40×10^6 cells/ml.

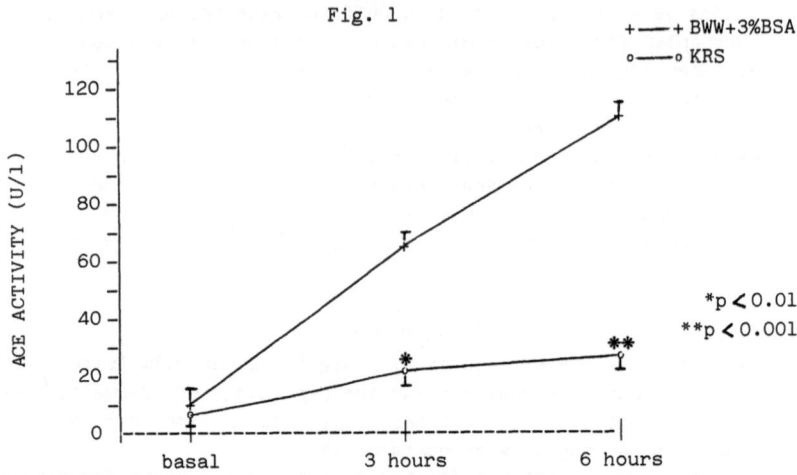

Fig. 1

ACE release by human spermatozoa incubated for 6 hours either in KRS or BWW. Each experiment was performed using samples of 40×10^{6} cells/ml.

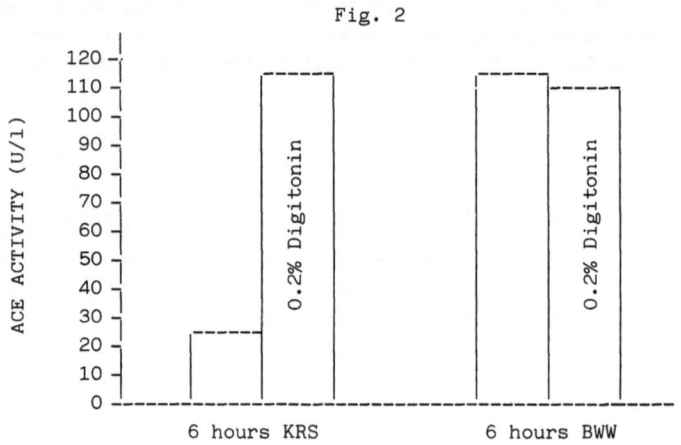

Fig. 2

Effects of 0.2% Digitonin on ACE release by spermatozoa incubated for 6 hours either in KRS or BWW. Each experiment was performed using aliquots of 40×10^{6} cells/ml.

Fig. 3

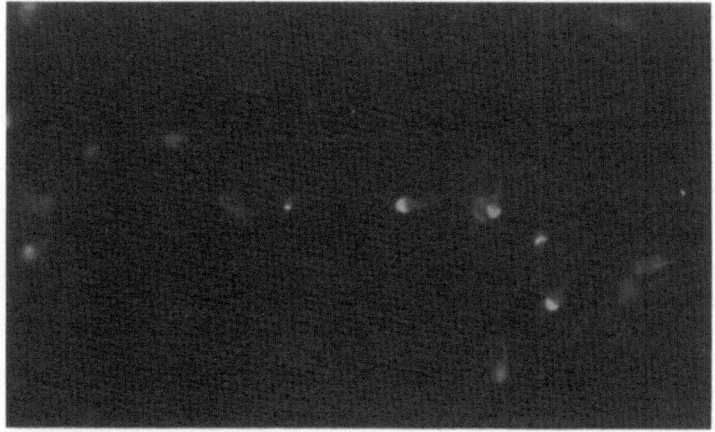

Immunofluorescent staining of human spermatozoa: the presence of the acrosome is revealed by the bright specific fluorescence, restricted at acrosomal cap level.

DISCUSSION

The disruption of ejaculated human sperm membrane by treatment with Digitonin and Triton X-100, substances known to exert a breakdown action on biological membranes, causes an important Angiotensin-Converting-Enzyme (ACE) release, demonstrating that ACE is contained within human spermatozoa.

ACE cleaves the decapeptide Angiotensin I to the octapeptide Angiotensin II, by removing a dipeptide from the carboxyl-terminus of Angiotensin I. Human ovarian follicular fluid contains high renin-like and angiogenic activity (9,10), therefore it is likely that spermatozoal ACE may be released within the female reproductive tract, during capacitation and acrosome reaction. These processes involve such important biochemical and morphological modifications of sperm membrane, so that the release of ACE may occur.

We were able to detect a high angiotensin-converting activity during incubation of human spermatozoa in capacitating medium, while this activity was slightly detectable when spermatozoa were incubated in non-capacitating conditions. The release of ACE seems not to be related to acrosome reaction, since it is maximal during 6 hours' incubation in BWW, when only 29% of spermatozoa show acrosome loss. On the other hand, ACE determinations in these conditions were similar to those obtained after Triton X-100 or Digitonin treatment. Therefore we believe that not acrosomes, but cytoplasmic residues, express angiotensin-converting activity.

The mechanism whereby the spermatozoal particulate ACE becomes soluble is unknown, but the release of ACE from endothelial cells provides a precedent (11) and we suggest that the release of ACE may be induced by the cellular metabolic activation occurring during the capacitation process.

CONCLUSIONS

The physiological significance of our findings is not clear, but we hypothesize that ACE could interact with the follicular renin-angiotensin system. Angiotensin-converting activity is maximal at alkaline pH and at a temperature of 37°C and the follicular fluid at periovulatory period provides these conditions: therefore we suggest that the physiological role of spermatozoal ACE may be played at female genital tract level.

REFERENCES

1. Holbrugger G, Pschorr J, Dahlheim H. Angiotensin I converting enzyme in the ejaculate of fertile and infertile men. Fertil Steril 1984: 41; 324-26.

2. Depierre D, Bargetzi JP, Roth M. Dypeptidyl carboxypeptidase from human seminal plasma. Biochim Biophys Acta 1978: 523; 469-73.

3. Vahna-Perttula T, Mather JP, Bardin CW, Moss SB, Bellvè AR. Localization of the Angiotensin-Converting- Enzyme Activity in Testis and Epididymis. Biol Reprod 1985: 33; 870-75.

4. Strittmatter SM, Thiele EA, DeSouza EB, Snyder SH. Angiotensin-Converting Enzyme in the Testis and Epididymis: Differential development and Pituitary Regulation of Isozymes. Endocrinology 1985: 117; 1374-77.

5. Strittmatter SM, Snyder SH. Angiotensin-Converting Enzyme in the Male Rat Reproductive System: Autoradiographic Visualization with (^{3}H)-Captopril. Endocrinology 1984: 115; 2332-34.

6. Holbrugger G, Schweisfurth H, Dahlheim H. Angiotensin I converting enzyme in rat testis, epididymis and vas deferens under different conditions. J Reprod Fertil 1982: 65; 97-100.

7. Cushman DW, Cheung HS. Angiotensin I converting enzyme in tissues of the rat. Biochim Biophys Acta 1971: 250; 261-65.

8. Wolf DP, Boldt J, Byrd W, Bechtol KB. Acrosomal Status Evaluation in Human Ejaculated Sperm with Monoclonal Antibodies. Biol Reprod 1985: 32; 1157-63.

9. Culler MD, Tarlatzis BC, Lightman A, Fernandez LA, DeClerney AH, Negro-Vilar AF, Naftolin F. Angiotensin II-like immunoreactivity in human ovarian follicular fluid. J Clin Endocrinol Metab 1986: 62; 613-14.

10. Fernandez LA, Tarlatzis BC, Rzasa PJ, Carida VJ, Laufer N, Negro-Vilar AF, DeClerney AH, Naftolin F. Renin-like activity in ovarian follicular fluid. Fertil Steril 1985: 44; 219-22.

11. Ching SF, Hayes LW, Slakey LL. Angiotensin Converting Enzyme in cultured endothelial cells and growth medium. Relationship to enzyme from kidney and plasma. Biochim Biophys Acta 1981: 657; 222-27.

Acrosin in human spermatozoa.
An immunohistochemical study

S. FRANCAVILLA, B. BRUNO, G. POCCIA, F. FRAN-CAVILLA, R. SANTUCCI, G. PROPERZI, V. SAN-TIEMMA, A. FABBRINI

Department of Internal Medicine, University of L'Aquila, L'Aquila, Italy

INTRODUCTION

A physiological barrier at level of cervix, uterine lumen, and Fallopian tubes, selects morphologically normal spermatozoa (1). Spermatozoa with defects of tail and mid-piece are excluded from reaching the site of fertilization, due to the impairment of their motility(2). Spermatozoa with abnormal head forms but no other sign of abnormalities, may reach in vivo, though in a small number, the uterus and oviduct (2), hence they are potentially able to fertilize the egg. With the object to analyze the potential intrinsic fertility of abnormally headed sperm in man, we evaluated by indirect immunofluorescence (IIF), the presence of acrosin in the heads of different morphologic types of ejaculated sperm. Acrosin is a proteolytic acrosomial enzyme which is required for fertilization of the egg (3), so its presence in the sperm head may give some information on the potential fertility of different forms of ejaculated spermatozoa.

MATERIALS AND METHODS

The data were obtained from 68 subjects randomly chosen among husbands of infertile couples, which were submitted to semen analysis following standard procedures (4). Acrosin was detected on fixed smears of ejaculated sperm by IIF according to Flörke-Gerloff et al. (5), by use of a specific antiserum obtained in rabbits against boar acrosin (6), gift of by dr. Müller-Esterl from University of Munchen. The morphology and occurrence of fluorescence over the anterior region of the head were recorded at the same time for each of 150 consecutive spermatozoa from each patient.

RESULTS

Analysis of sperm morphology revealed an abnormal shape in 50% of spermatozoa. Abnormal forms included defects of the head in 52% of all abnormal cells, while defects of the mid-piece and of the tail appeared, respectively, in 27% and 21% of the abnormal spermatozoa. In any ejaculate which displayed abnormal headed sperm, cells with amorphous he-

221

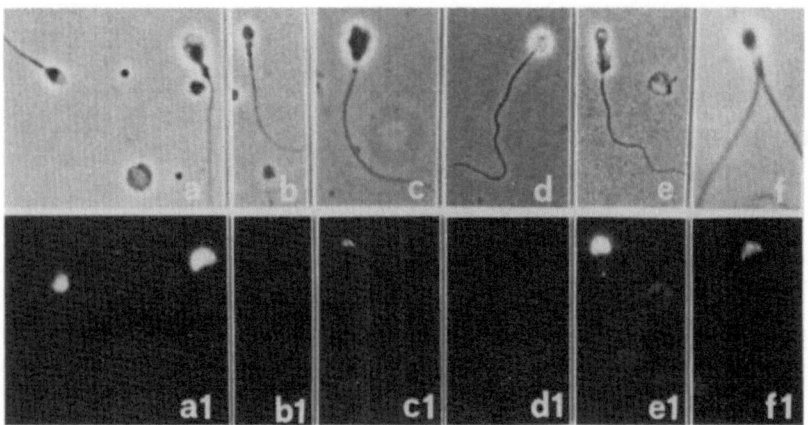

Figure 1. Localization by IIF of Anti Acrosin Antibodies in different forms of ejaculated spermatozoa. a) normal (left) and large oval head (right); b) small oval head; c) amorphous head; d) round head; e) cytoplasmic remnant in the mid-piece in a normal headed sperm; f) double tail in a normal headed sperm. Acrosin is absent in small and round headed sperm, and is restricted to a small spot in the amorphous headed sperm.
a) – f) phase contrast, x 480. a_1) – f_1) IIF, x 480.

Table I. Percent of Acrosin positive (+) and negative (–) cells in each sperm morphologic type

Sperm forms	Acrosin(+) mean %	Acrosin(–) mean %	± SD	t*	P
Normal forms	84	16	11	13.2	a
Mid-piece defects	87	13	17	8.7	a
Tail defects	85	15	27	7.3	a
Head defects	9	91	10	–13.4	a
– Amorphous	4	96	8	–10.9	a
– Tapered	10	90	21	– 6.8	a
– Round	4	96	11	– 7.3	a
– Small oval	12	88	23	– 3.3	a
– Large oval	81	19	30	4.1	a

a: $p < 0.001$

*: The student t test was applied to the differences between the mean percent of Acrosin positive and negative cells in each subject by previous transformation of data in logit.

ads were constantly present while all other defects of the head including large and small oval, tapered, and round forms were less numerous.
The IIF test for acrosin showed that the antigen had a homogeneous cup-
shaped distribution in the acrosomial region of the normal oval or large headed sperm (Fig. 1a), while the fluorescence was absent in most
of the small, tapered, amorphous and round heads (Fig. 1b-d). Scattered abnormal headed sperm showed small spotty deposits of the antigen
in the anterior region of the head (Fig. 1c). Most of the spermatozoa
with defects of the mid-piece and of the tail, but with a normal oval
head, showed a regular distribution of the antigen (Fig. 1e,f). All
categories of sperm morphology were then compared in each patient over
the whole population of 68 subjects for the frequency of acrosin positive and negative cells in each category (Tab. I). For each morphology
type, a significant difference resulted between the mean percent of
the acrosin positive and negative cells (p < 0.001). A mean of more than
80% of spermatozoa resulted positive at IIF test for acrosin in the categories of normal or large headed cells, and in the categories with
mid-piece and tail defects. On the contrary, the antigen was present
in a mean of less than 12% of amorphous, round, tap ered, and small he
aded spermatozoa. The analysis of variance with multiple confronts (7)
revealed a significant difference (p < 0.005) between the two groups of
categories of sperm morphology, while no difference resulted between
categories included in each of the two groups.

DISCUSSION

Acrosin is a proteolytic enzyme, located as a zymogen proacrosin in
the acrosome of the spermatozoon (8), which is required for the fertilization of the egg (3).
The present study shows that this enzyme is absent not only in round
headed sperm as already documented (5), but in all morphologic types
of spermatozoa with atypical heads, with the exception of large oval
forms. Defects of the mid-piece, and of the tail in otherwise normal
headed sperm are on the contrary associated to a regular cup-shaped
appearance of acrosin. This suggests that spermatozoa with isolated
head defects, though potentially able to penetrate the cervical mucus
and to reach the Fallopian tubes (2), cannot fertilize eggs. The abnormal heads which displayed a positive reactivity for acrosin showed an
abnormal staining pattern of the antigen which appeared in fact as a single or multiple fluorescent spots. The ultrastructural analysis gave
consistent results with the immunohistochemical findings showing that
abnormal headed sperms had always an abnormal or absent acrosome (unpublished).
This suggests that factors which control the normal shape of the spermatozoon head are tightly related to those which are responsible for
a normal differentiation of the acrosome and synthesis of its enzymatic content.
Sperm shape is defined by the genotype of the spermatogenic cells (9).
Comparative abservations in different species including mammals, birds
insects and annelids have suggested that the intrinsic morphogenetic
forces which model the sperm head may reside within the nucleoplasm
and in the substance of acrosome (10). Therefore, an inherent absent
or abnormal synthesis of acrosin could contribute to an abnormal development of the acrosome and of the sperm head. Abnormalities of the
sperm head are largely the result of genetic defects of primary sper-

matocytes and spermatogonia (11), which become manifest in the haploid phase of maturation of germ cells.

In this light, the lack of acrosin in abnormally headed, but alive and potentially motile sperm, prevents fertilization of the egg by spermatozoa which carry genetic defects.

REFERENCES

1. Bergman P. Spermigration and its relation to the morphology and motility of spermatozoa. Int.J.Fertil. 1955: 1;45-55.

2. Mortimer D., Leslie E.E., Kelly R.W. and Templeton A.A. Morphological selection of human spermatozoa in vivo and in vitro. J.Reprod. Fertil. 1982: 64;391-399.

3. Zaneveld L.J.D. Sperm enzyme inhibitors as antifertility agents.In: Human Semen and Fertility Regulation in Man. (1976) p. 570 Ed. E.S. E. Hafez, Mosby Company, St.Louis.

4. W. H. O. Manual for the investigation and diagnosis of the infertile couple. (1979) Study No. 78923. Special Program in research, development, and training in human reproduction, Géléve.

5. Flörke-Gerloff S., Topfer-Petersen E., Müller-Esterl W., Mansouri A., Schatz R., Schirren C., Schill W., and Engel W. Biochemical and genetic investigation of round-headed spermatozoa in infertile men including two brothers and their father. (1984) Andrologia 16:187-202.

6. Müller-Esterl W., Küpfer S., and Fritz H. Purification and properties of boar acrosin. (1980) Hoppe-Seyler's Z. Physiol.Chem. 361: 1811-1821.

7. Armitage P. Statistical methods in medical research. (1980) p. 207. New York: John Wiley.

8. Polakoski K.L., and Parrish R.F. Boar acrosin: purification and preliminary activation studies of proacrosin isolated from ejaculated boar sperm. (1977) J. Biol. Chem. 252:1888-1894.

9. Beatty. R.A. The genetics of size and shape of spermatozoan organelles. In: The Genetics of the Spermatozoan, Proceedings of an International Symposium, Edinburgh, Scotland 1971. Eds. R.A. Beatty and S. Gluecksohn-Waelesch. (1972) p. 97 Department of Genetics, University of Edinburgh.

10. Fawcett D.W., Anderson W.A., and Phillips D.M. Morphogenetic factors influencing the shape of the sperm head. (1971) Dev.Biol. 26: 220-251.

11. Burgoyne P.S. Sperm phenotype and its relationship to somatic and germ line genotype: a study using mouse aggregation chimeras. Dev. Biol. 44:63-76.

Evaluation of zinc content in human spermatozoa

M. BARTELLONI, D. CANALE, P.M. GIORGI,
P. TURCHI, P. MESCHINI, P. GIANNOTTI,
G.F. MENCHINI FABRIS

Postgraduate School of Andrology, 1st Medical Clinic, University of Pisa Medical School, Pisa, Italy

Introduction

During the last years the biochemical components of seminal plasma have been the subject of many researches for their influence on the fecundating potential of semen and their role of markers of accessory glands function. Zinc, mainly secreted from the prostate, has been studied in order to verify its possible role as marker of prostate activity (1). Zn has a stabilizing effect on membranes, where it is bound to lipoproteins. The influence of this ion on testis function and its role on the energy metabolism of spermatozoa have been pointed out in previous studies (2). We have already reported data concerning the Zn level in the seminal plasma of normal and infertile patients using a new spectrophotometric technique (3). In this work we report the first data of intraspermatozoal Zn assessed by the above-mentioned technique.

Materials and Methods

Zn determination were carried out in the spermatozoa of 37 patients referring to our Center to test and evaluate their fertility potential. They were divided into groups as follows: oligoasthenospermic (n=7); varicocele (17); hyperprolactinemia (2); prostatitis (2); normal (control group) (9). The seminal plasma has been collected by masturbation after a 3-4 day abstinence.

Semen analysis was carried out according to the criteria suggested by W.H.O. (5).

The preparation of sperm from seminal plasma was obtai-

Table 1. Mean intraspermatozoal Zinc content (± SE) in various pathologies examined

PATIENTS	CASES	INTRASPERMATOZOAL ZINC CONTENT*
CONTROL	9	13.0 ± 1.4
OLIGOASTHENOSP.	7	42.0 ± 6.9
VARICOCELE	17	34.3 ± 5.1
HYPERPROLACT.	2	74.0
PROSTATITIC	2	32.0

* mcg/10^8 cell.

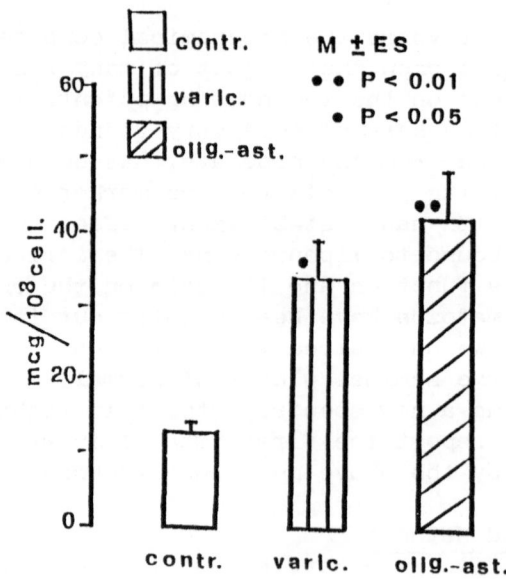

Fig.1. Correlation between mean (± SE) intraspermatozoal Zn content in control group and varicocele and oligo-asthenospermic patients.

ned by centrifugation (5', 3000 rpm), and subsequent washing with a saline solution. The sperm suspension is then
treated as the whole semen in our previous work (3-4).

Results
The mean value of intracellular Zn in the group of 9 normo
spermic subjects (seminal plasma volume between 2-4 ml,
total number of spermatozoa > 40 mil/ml, > 50% of forward
progressive motility, > 70% of normal morphology) is 13.0
($\pm$ 1.4 SE) mcg/10 cells (Table I). In the same table the
values of intraspermatozoal Zn in oligoasthenospermic patients, varicocele, prostatitic and hyperprolactinemic
patients are shown. The correlations between the Zn levels
in the control group and the groups of patients with varicocele and oligoasthenospermic semen are reported in figure
1.

Discussion
In this work, the Authors have shown firstly that the colo
rimetric technique for determination of intraspermatozoal
Zn can represent a valuable support in studying the metabolism of this metal ion. Besides, this technique is rapid, easily reproducible and economical.
The findings show that in pathological conditions such
as oligoasthenospermia, varicocele and prostatic inflammation , the intraspermatozoal Zn content increase significantly ($p < 0.05$ for the varicocele group and $p < 0.01$
for the oligoasthespermic one, compared with the control
group). These high Zn concentrations could inhibit a sufficient release of the ion during the travel through the
female genital tract. Such condition would determine difficulties in the chromatin decondensation which occour
immediately and concomitantly with the sperm membrane disintegration upon sperm penetration into the ooplasm(6-9).
The values found in sperm from hyperprolactinemic patients confirm the positive correlation between intraspermatozoal Zn content and prolactin, shown also by
other Authors (10). However, this finding needs further
studies.

References

1. Eliasson R, Lindholmer C. Zinc in human seminal plasma.
   Andrologia 1971: 3;147-53.
2. Eliasson R, Johnsen O,Lindholmer C. Effect of Zinc on
   human sperm respiration. Life Sci 1971: (1) 10; 1317-
   21.

3. Bartelloni M, Negroni A, Meschini P, Olivieri L, Voliani S, Canale D, Menchini Fabris GF. Lo Zinco nelle dispermie. In: Aggiornamenti in Andrologia Chirurgica. D'Ottavio G, Pozza D. Eds. Roma. Acta Medica 1986: 35-37.
4. Lampugnani L, Maccheroni M. Rapid colorimetric of Zinc in seminal fluid. Clin Chem 1984: 30; 1366-68.
5. Laboratory Manual for the examination of human semen and semen-cervical mucus interaction. W.H.O. Report. Singapore. Press Concern 1980.
6. Delgado NM, Huacuta L, Pancardo R, Rosado A. Modification of human sperm metabolism by the induced release of intracellular Zinc. Life Sci 1975: 16; 1483-88.
7. Kvist V, Eliasson R. Influence of seminal plasma on the chromatin stability of ejaculated human spermatozoa. Int J Androl 1980: 3; 130-42.
8. Kvist V. Reversible inhibition of nuclear chromatin decondensation (nco) ability of human spermatozoa induced by prostatic fluid. Acta Physiol Scand 1980c: 109; 73-78.
9. Björndhal L, Kvist V. Loss of an intrinsic capacity for human sperm chromatin decondensation. Acta Physiol Scand 1985: 124; 189-94.
10. Leake A, Chisholm GD, Habib FK. Interaction between prolactin and Zinc in the human prostate gland. J Endocr 1984: 102; 73-76.

The use of echotomography in the diagnosis of male infertility

G.F. MENCHINI FABRIS, M. SARTESCHI, R. PAOLI, P.M. GIORGI, G. ESPOSITO, D. CANALE

Postgraduate School of Andrology and Andrology Center, 1st Medical Clinic, University of Pisa Medical School, Pisa, Italy

The ejaculate is a suspension of cells (spermatozoa) in a fluid medium, made of secretions coming from the various glands accessory to the secretory genital pathways. This equilibrium, both qualitative and quantitative, allows an adequate development of sperm movements and, therefore, of fecundation capacity (1).

One of the most frequent causes of male infertility is represented, nowadays, by the genital inflammations. They induce a reduction of the fertility potential mainly by determining an alteration of accessory gland secretion. As a matter of fact, the inflammatory processes, either acute or chronic, are able to deeply modify the functions of epididymis, ampullae, prostate, vesicles and glands of Cowper -Littré (2).

To diagnose this kind of pathology, the available tools have long been, beside sperm analysis and culture, only the data obtainable with the physical examination. A great advance has been accomplished with the introduction of deferento-vesiculo-graphy (DVG). This technique was first performed by Puigvert in 1939, but has entered the diagnostic routine only in the last 15 years. In 1974, Boreau, in his "Images of the seminal tracts", stated that "...hormonal investigations carried out together with radiographic examination of the seminal ducts reflects faithfully the endocrine activity of the subject, more than his sexual capacities. .... It is not possible to obtain normal sperm unless the ampullae and the seminal vesicles are healthy." This technique does really permit a good morphological investigation of vas deferens, ampullae, seminal vesicles, ejaculatory ducts and epididymis as we have experienced in these years in collaboration with the General Surgery Dept. of the Pisa University (3). But this method implies a surgi cal act, the introduction of a contrast medium, the irradia

tion of the gonads, the possible occurrence of stenosis or
substenosis of vasa. A further progress in diagnosis of the
inflammatory pathologies of the male genital tracts has been
performed with the use of echotomography, which we have extensively and routinary used in these last 3 years. This
technique provides a good morphological and echostructural
evaluation of the secretory pathways and of the accessory
glands, using adequate probes and " hands ".
The use of ultrasound, first limited to noticeable external
genital and prostatic pathology, has been subsequently extended to the diagnosis of " minor " but not less important for the andrologist, alterations (4). Particularly,
echotomography has revealed great and unexpected possibilities of diagnosis in male genital inflammatory conditions,
in which an invasive method such as DVG could have undesi
rable effects. Moreover, the possibility of repeating it
with ease permits the use of this method in the follow-up of
any therapy (5).
In this paper will be presented a study on infertile patients affected by genital inflammations in which are compa
red the echographic with the seminal patterns.

SELECTION OF CASES

In figure 1 is shown a cyst right caput epididymis with a
spermatocele. This echographic finding was accompanied by

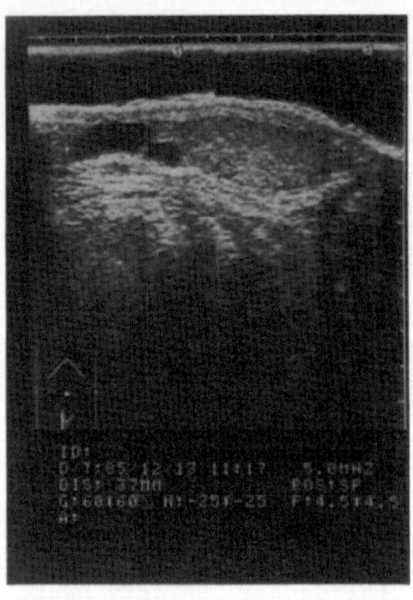

Fig. 1:
cyst right caput epididymis
with a spermatocele.

a seminal pattern characterized by asthenospermia, a high number of immature sperm and reduction of seminal carnitine.

Figure 2 shows an acute epididymitis with increased size of the gland and presence of hypoechogenic and dyshomogeneous areas within it associated with a hydrocele. The seminal fluid was similar to the previous one, with a greater number of leukocytes and agglutinates a condition of chronic epididymitis with calcified nodules associated with vaginalitis (with an adhering calcium nodule) is represented in fig. 3. In this case semen analysis revealed a complete asthenospermia with teratospermia.

In fig. 4 can be observed in trans-rectal median and 30° scan the typical pyramidal shape of a normal prostate with the base on the bladder floor and the apex at the uro-genital diaphragm (R = rectum; V = bladder; P = prostate; FD = Denonvillers' fascia; SP = pubic symphisis; DU = urogenital diaphragm).

In fig. 5 a chronic prostatitis with clear glandular calcium depositions, but without modifications of glandular morphology and margins is shown. The semen analysis revealed an alteration of sperm motility and morphology (high incidence of head-tail deconnections and bent-tails) and abnormalities of liquefaction and viscosity. An image of normal seminal vesicles in trans-rectal linear 60° scansion is

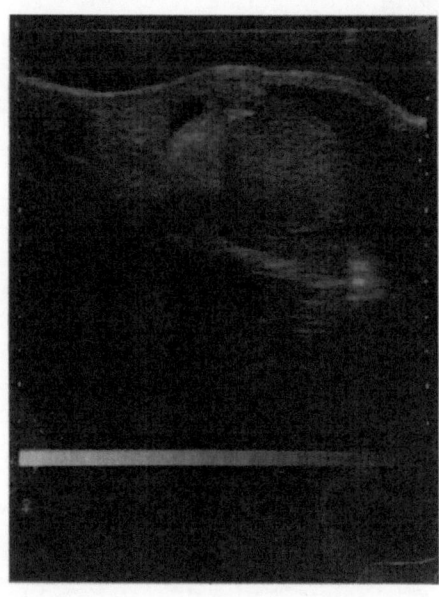

Fig. 2.

acute epididymitis.

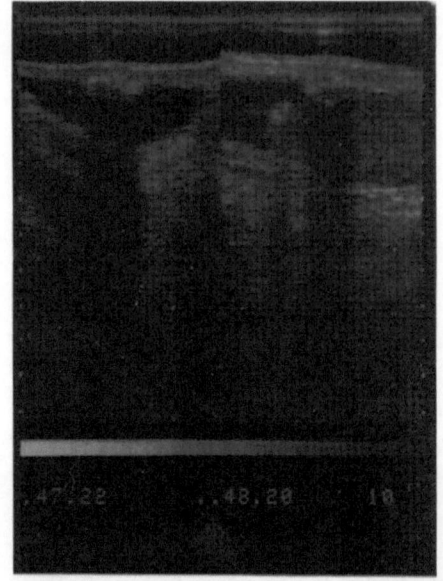

Fig. 3.
chronic epididymitis.

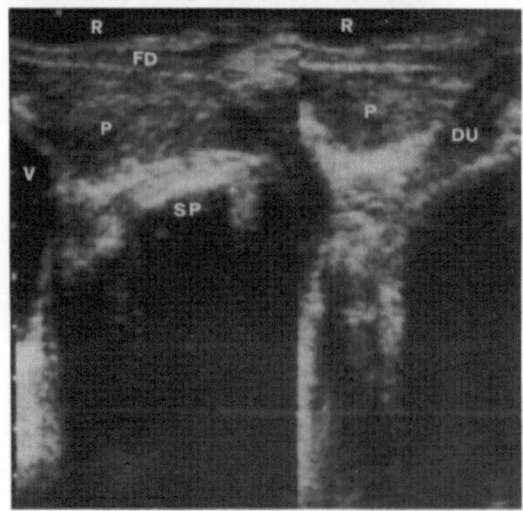

Fig. 4.
normal prostate.

shown in fig. 6: they have a tubular profile, clear margins and homogeneous echogenicity. During acute vesiculitis (Fig 7), they can appear enlarged, "barrel-shape-" (mainly the right one), quite hypoechogenic (the right one thoroughly unechogenic), with thickened walls and painful at the aimed pression with the probe. The seminal fluid had an increased volume, basic pH, with a relevant number of leukocytes and agglutinates.

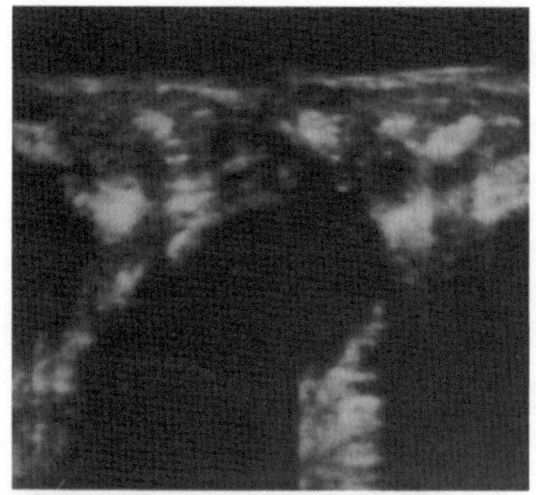

Fig. 5:
chronic prostatitis.

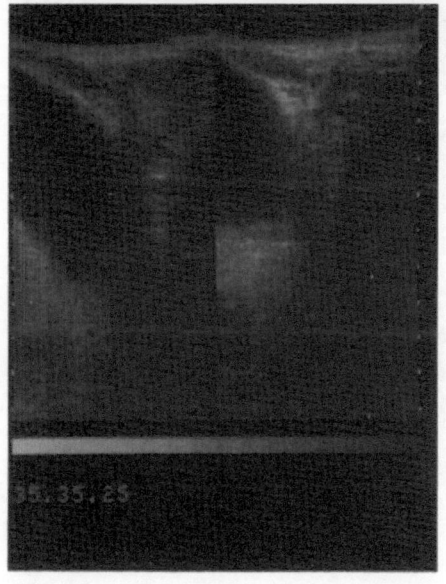

Fig. 6:
normal seminal vesicles.

In fig. 8 and 9 are shown two cases of subacute and chronic vesiculitis, respectively. In the first figure the vesicles are still enlarged, whilst in the second they have a reduced volume. In both they are dyshomogeneous and with thickened walls. Semen volume tended to become normal or re duced (as in the second case); there were still numerous leukocytes and sperm agglutinates; motility and viability were strongly reduced.

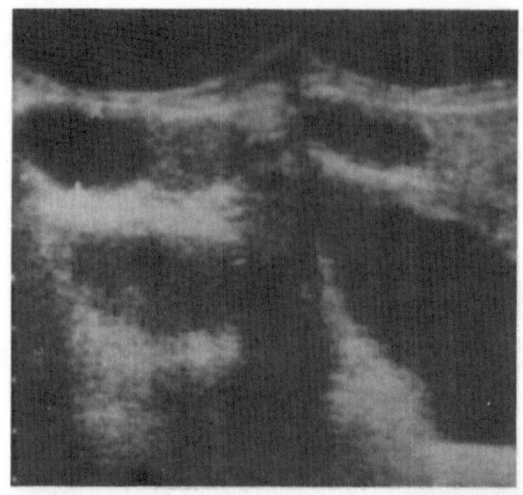

Fig. 7:
acute vesiculitis.

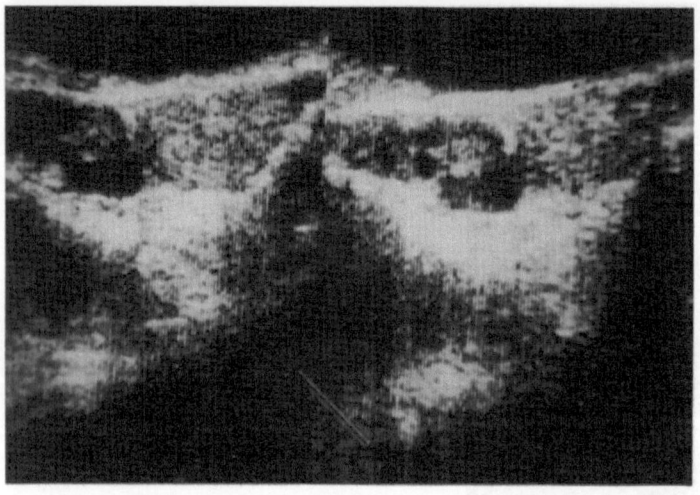

Fig. 8: subacute vesiculitis.

Fig. 10 shows a rare case of Cowper's glands hypertrophy. They appear distinctly enlarged (the left one has a diameter of 1.5 cm) and moderately hypoechogenic. The seminal fluid had always a strongly increased viscosity, a delayed liquefaction, a normal sperm count with normal morphology and a reduced viability.

Lastly, a sclerosis of ejaculatory ducts associated with calcified areas in the posterior region of the prosta-

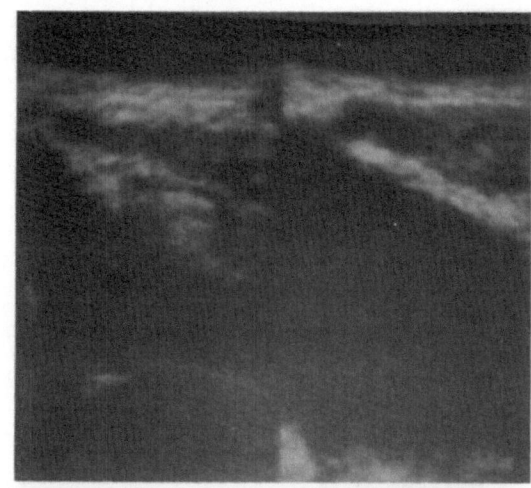

Fig. 9:
chronic vesiculitis.

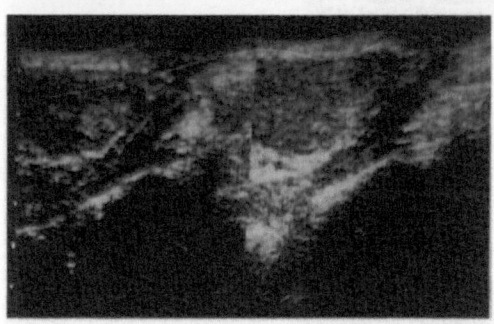

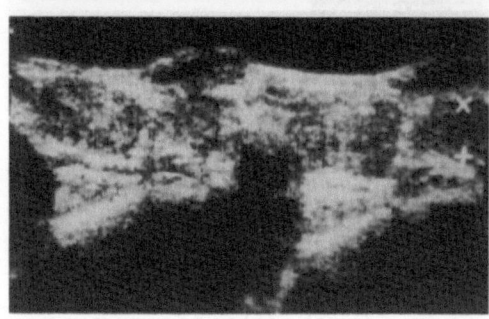

Fig.10:
Cowper's glands hypertrophy.

te gland can be seen in fig. 11. It was from a patient with
operated bilateral varicocele and recurrent prostatitis.
This patient had suffered from frequent hemospermias associated with reduced sperm motility and a high number of abnormal forms.

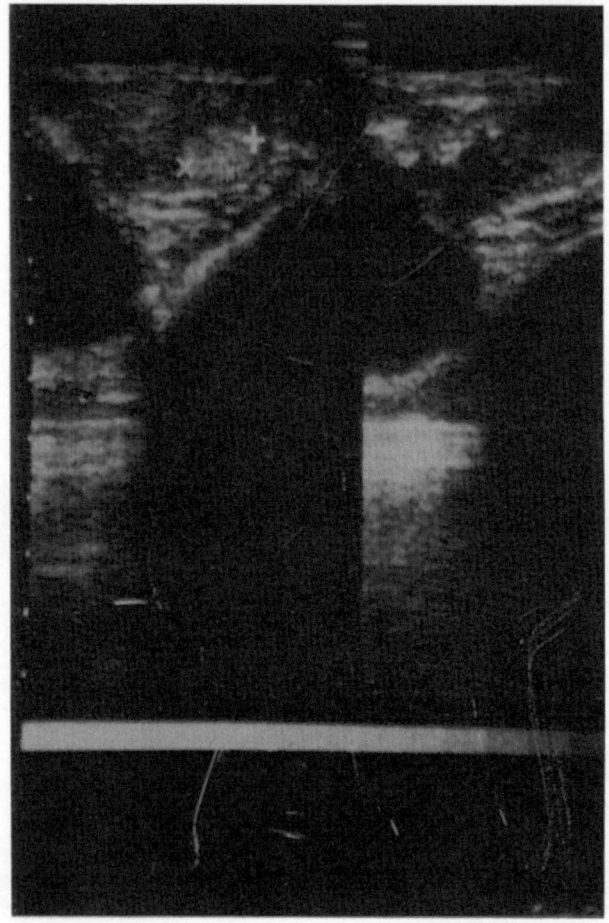

Fig. 11: chronic prostatitis

CONCLUSIONS

In this short article, we wanted to draw   attention to
the importance of echotomography in diagnosis of male infer
tility. Analogously to what occurs in obstetrics and gynecology, it should be correct that an ultrasound investigation is carried along with the other diagnostic procedures
in andrological routine, such as hormonal assessment, semen
analysis, etc.

In conclusion, the use of ultrasound should be conside
red a valuable tool for the identification, evaluation and
control of the follow-up of a pathology, in which the dia-

gnostic procedure, beside DVG, were almost " blind "as far
as concerns objective data.

REFERENCES

1) Menchini-Fabris,G.F., Voliani, S., Canale, D., &
   Olivieri, L. Sperm analysis in the infertile male.
   In: Andrology: Male Fertility and Sterility, eds.
   Negro-Vilar, A., Martini, L., Lucena, E. New York,
   Academic Press, 1986: 341-357.
2) Cavirani, C., Bianchi, B., & Menchini-Fabris, G.F.
   Flogosi genitali maschili non veneree e patologia
   riproduttiva. In: Andrologia,Clinica e Sociale, ed.
   Menchini-Fabris, G.F. Roma, CIC, 1985; 195-211.
3) Giannotti,P. & Menchini-Fabris, G.F. Some thoughts
   on the use of deferentovesiculography(DVG). In:
   Male Fertility and Sterility, eds. Mancini, R.E. and
   Martini, L. New York, Academic Press, 1974: 539-544.
4) Pintauro, W.L. et al. The use of ultrasound for
   evaluating subacute unilateral scrotal swelling. J.
   Urol 1985 May; 133 (5): 799-802.
5) Fujino, A. et al. Transrectal ultrasonography for
   prostatic cancer: its value in staging and monitoring
   the response to radiotherapy and chemoterapy. J. Urol
   1985 May: 133 (5): 806-810.

Morphological examples of the antisperm immune reaction

F. DONDERO, A. LENZI

Laboratory of Immunology of Reproduction and Seminology, 5th Medical Clinic, University of Rome "La Sapienza", Rome, Italy

INTRODUCTION

Antisperm immune reaction is at present considered one of the possible causes of pathology in infertile couples (1). Morphological aspects of this pathology can be considered from two different perspectives. The first studies the characteristic morphological patterns consequent to the presence, in vivo, of antisperm antibodies in male and female genital secretions. The second studies the evidence of the antisperm immune reactions artificially induced during laboratory tests for antisperm antibodies detection.

In this short paper we would like to report two typical examples of the above-mentioned situations.

EXAMPLE No. 1. If sperm agglutinating antibodies are present in the semen of patients, they can induce the formation of areas of sperm agglutination (2). Semen analysis, correctly carried out, could then indicate the presence of areas characterized by vital spermatozoa agglutinated in small clusters that tend to become larger during the observation and to keep a vibratory motility (Figs. 1, 2, 3, 4).

EXAMPLE No. 2. The more recent test for antisperm antibody detection, Immunobead Test (3), presents interesting morphological indications of the presence and localization of antisperm specific immunoglobulins. The markers are represented by spherules coated with anti-human Ig of various classes that are analyzed with a light microscope. Here we report some evidence to Scanning Electron Microscope (S.E.M.) (Figs. 5, 6, 7, 8, 9, 10).

239

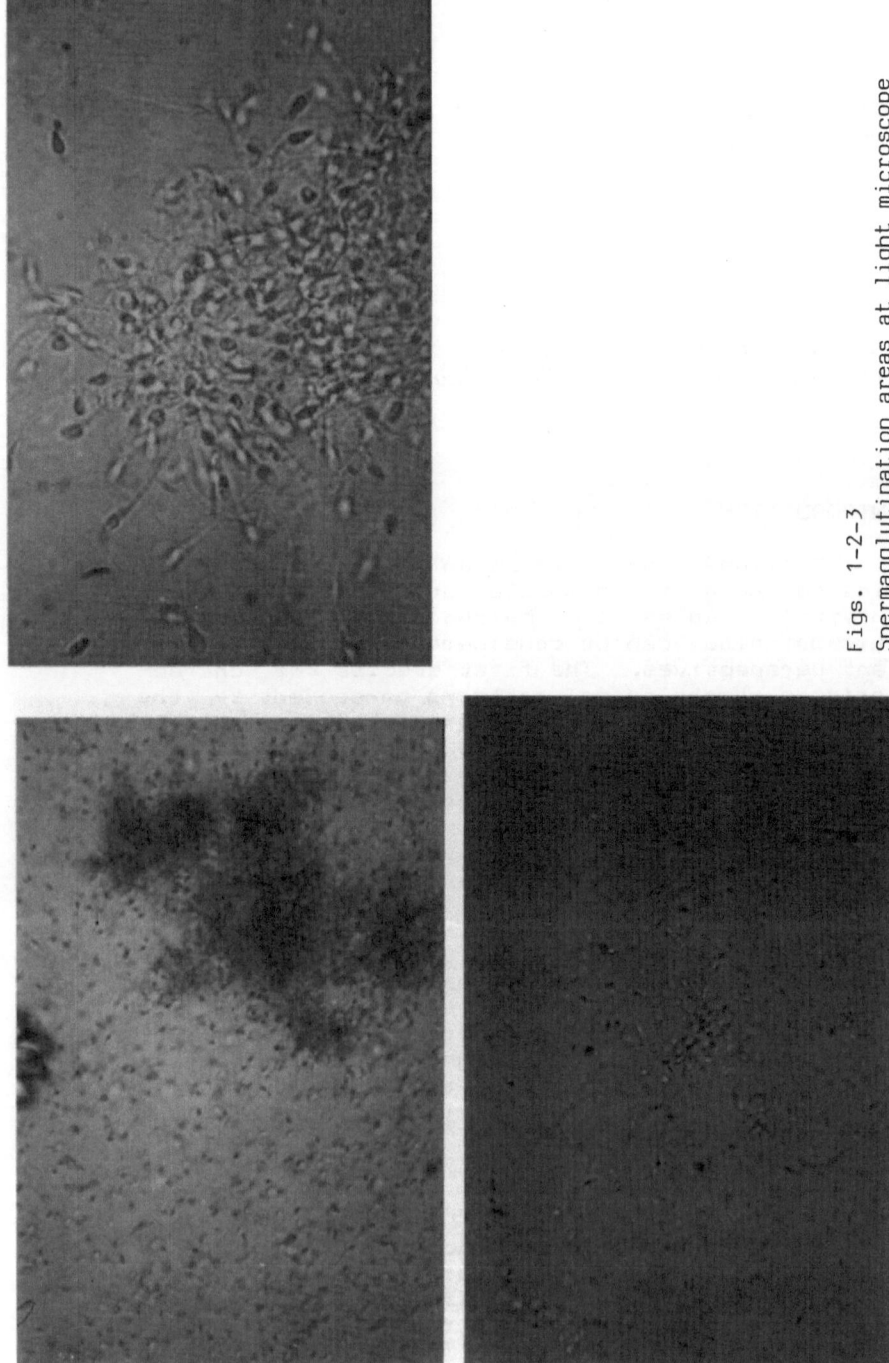

Figs. 1-2-3
Spermagglutination areas at light microscope

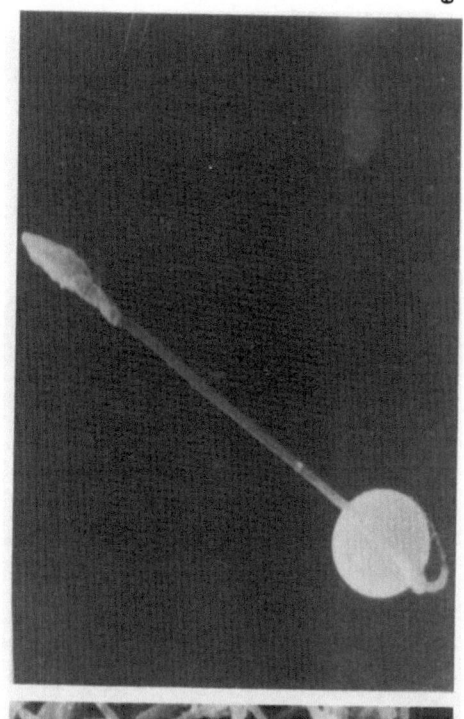

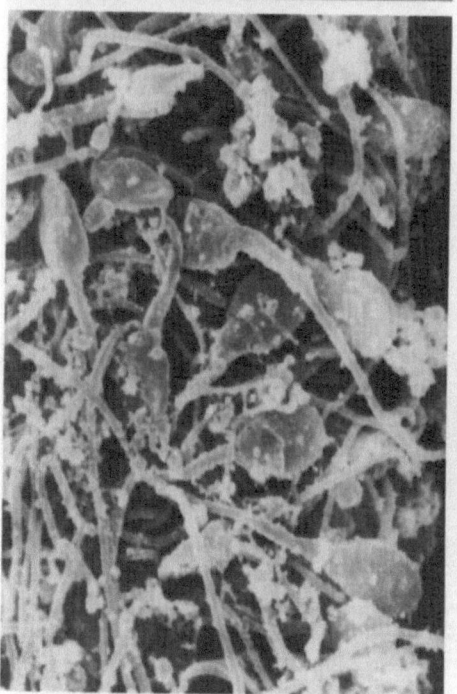

Fig. 4
Spermagglutination areas at S.E.M.

Fig. 5
Immunobead Test at S.E.M.: negative result

Fig. 6
Immunobead Test at S.E.M.: Positive result (tail)

241

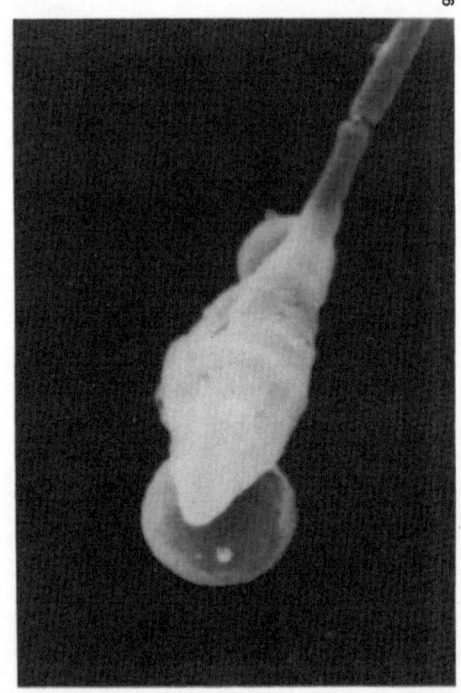

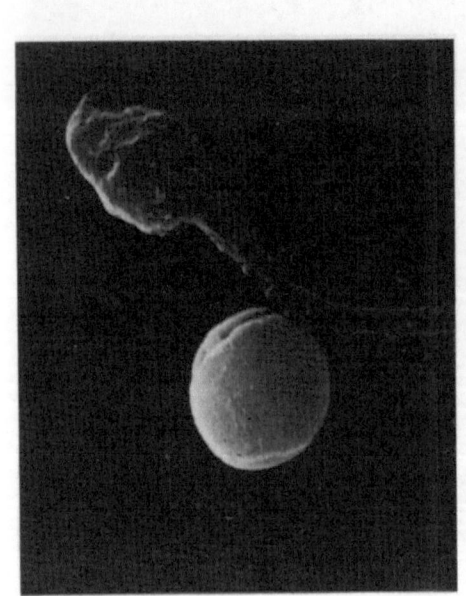

7

8

9

Figs. 7-8
Immunobead Test at S.E.M.: positive result (head)
Fig. 9
Immunobead test at S.E.M.: Positive result (head-nack)

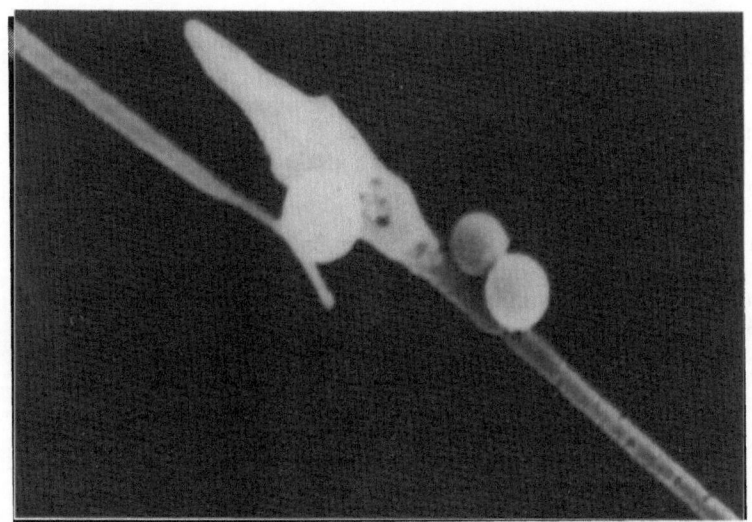

FIGURE 10. Immunobead Test at S.E.M.: positive result (head-neck-tail).

CONCLUSIONS

The above few examples demonstrated that morphological studies and techniques are essential in all steps of the diagnosis of immunological human infertility.

ACKNOWLEDGEMENTS
The Authors thank Dr Alberto Caggiati (Dept. Anatomy — University of Rome "La Sapienza", Dir. Prof. PM. Motta) and Drs Filiberto Claroni and Loredana Gandini (5th Medical Clinic — University of Rome "La Sapienza", Dir. Prof. C. Conti) for their kind collaboration.

REFERENCES

1) Current Trends in Immunology of Human Reproduction, eds. Shulman S, Dondero F. J Immunol Immunopharmacol 1986, 6, 2.

2) Dondero F, Mazzilli F, Giovenco P, Lenzi A, Isidori A. Morphological and biochemical patterns of semen before and after treatment in patients with infertility due to spermagglutinins. In: Recent Progress in Andrology, eds. Fabbrini A, Steinberger D. New York, Academic Press, 1978: 427:431.

3) Bronson RA, Cooper GW, Rosenfeld DL. Ability of antibody-bound sperm to penetrate zona-free hamster ova in vitro. Fertil Steril 1981:36, 778-783.

Cytology and cytogenetics of human semen: importance of the association of two different approaches in andrological diagnosis

F. MENCHINI FABRIS, L. OLIVIERI, G. ESPOSITO, F. ANDREINI, S. VOLIANI

Postgraduate School of Andrology, Clinica Medica I, University of Pisa, Italy

The increase in number of spermatogenetic cells in semen, or the presence of structurally abnormal ones (plurinucleated spermatocytes and spermatids) may be the result of pathological events, so that their identification and classification may be useful for the diagnosis of andrological diseases. The identification of different cell types is, at times, difficult in phase-contrast observation of fresh material: in these instances the classification must rely upon the size and the shape of the entire cell, or the size, shape and position of the nucleus, or the presence of cytoplasmatic granules. To have a good cell differentiation, seminal smears should be stained according to conventional cytological techniques, such as the May-Grunwald-Giemsa, the Papanicolau's or the Bryan-Leishman stain. Using these methods, immature germ cells can be differentiated from leukocytes and can be well-classified according to their maturative stage. Spermatogonia are rarely found in semen: their diameter ranges from 5 to 12 µ and the nucleus is about half the cell: these cells usually contain one or two nucleoli that may be seen resting on the edge of the nucleus. Primary spermatocytes are larger than spermatogonia, their cytoplasm is pale and their large nucleus is usually homogeneous, but occasionally chromatin threads are seen; binucleated elements are fairly common while plurinucleated ones suggest meiotic disorders. In semen, secondary spermatocytes are spherical, with diameter of 8-12 µ; their nucleus is very large, occupying most of the cell, and shows affinity for basic stains. Spermatids, in the ejaculate, show a well-marked nucleus which appears as an excentric compact sphere; at fresh observation, spermatids are often difficult to distinguish from leukocytes, but this distinction becomes easy in stained preparations.

The study of immature germ cells is not limited to their count and classification. Cytogenetic techniques are available for the visualization of meiotic stages directly from the semen, thus avoiding to perform testicular biopsy (1, 2). The analysis of meiosis, in addition to

245

to show the structure and behaviour of germinal chromosomes, makes us able to recognize if spermatogenetic breakdown is due to cytogenetical abnormalities. In these instances, low chiasma count, asymmetrical bivalents, multivalents, univalents, fragmentation and polyploidy may be found. The first prophase is the longest and most easily observable phase in meiosis: the paired chromosomes demonstrate a zipper-like con figuration, while X and Y chromosomes are compressed into a dark body, the so-called sex vesicle. Bivalents can be counted and identified from diakynesis to first metaphase, and during these stages it is possible to find polyploid cells; the X-Y bivalent appears as a long rod bivalent, but, sometimes, they appear as two univalents; at these stages is also possible to make the chiasma count. The chromosome in second metaphase are difficult to identify because of their fuzzy and twisted appearance (3).

Since the cytology and the cytogenetics of immature germ cells can be studied in semen as well as in testicular tissue, a valuable contribution to Andrologists is provided by the combination of these two dif ferent approaches. The main parameters which must be evaluated are: the spermatogenetic cells density and the germ cells/spermatozoa ratio; the maturative stage; the percent of polyploidy and multinucleated cells; pairing abnormalities (lack of sex-vesicle in prophase figures) and structural abnormalities in meiotic chromosomes, the rate of degeneration. By means of these parametres, testicular damage, excretive pathways suffering, reduced spermatogenetic activity and meiotic break down may be recognized. The study in semen can be repeated at will, thus it is possible to follow the progress of the patient and his response to treatment; it describes what happens in the entire testes and permits cytological and cytogenetical study on the same specimens; it is not thraumatic and gives quick results.

REFERENCES
1. Sperling K. and Kaden R. Meiotic studies on the ejaculated seminal fluid of humans with normal sperm count and oligozoospermia. Nature 1971; 232, 481-2.
2. Templado C., Marina S. and Egozcue J. Three cases of low chiasma frequency associated with infertility in man. Andrologia 1984; 8, 285-289.
3. Menchini Fabris F., Voliani S., Olivieri L. e Izzo P.L. Meiosis in infertile men. In: Oligozoospermia: Recent Progress in Andrology, Ed. G. Frajese. New York, Raven Press, 1981: 267-73.

Acrosomal status evaluation during capacitation of human sperm used for "Sperm Penetration Assay" (SPA)

F. FRANCAVILLA, S. FRANCAVILLA, B. BRUNO,
R. ROMANO, L. CASASANTA, R. SANTUCCI,
P. CATIGNANI, M. MARTINI

*Department of Internal Medicine, University of L'Aquila,
L'Aquila, Italy*

INTRODUCTION

The widespread agreement on the clinical usefulness of SPA is justified on the basis of the correlation which almost constantly has been reported between its results and fertility status (1). But the functional nemaspermic property which SPA can evaluate is not yet clear. Since only reactated sperm can actively penetrate zona-free hamster ova (2), SPA could evaluate sperm capacitation and acrosomal reaction and/or physiological events related to acrosome reaction. Recent reports seem to demonstrate that only a little percentage of sperm spontaneously exhibits acrosome reaction during "in vitro" capacitation in man (3,4). We evaluated acrosomal status during capacitation for SPA by immunofluorescence test using an antiserum against acrosin, and ultrastructural observations. We looked also for a relation between acrosomal loss and results of SPA.

MATERIALS AND METHODS

12 male partners with infertile marriages (TAB.I) and 6 fertile controls were studied.

SPA. Standard procedures were utilized (1). In order to achieve capacitation, motile sperm suspensions were obtained by swim-up procedure in BWW+3.5% HSA and then incubated at 37°C, 5% CO_2 for 8 hours.

Indirect Immunofluorescence (IIF). Drops of sperm suspensions before and after capacitation were smeared on glass slides, air dried, methanol-fixed for 30 min. and stored at -35°C. The smears were incubated in a moisture chamber for 60 min. with a specific antiserum raised in rabbits against purified boar acrosin, at a dilution of I:40. The antiserum resulted highly specific for acrosin and pro-acrosin of human sperm (5). After 3 washings in PBS, the smears were incubated for 30 min. with FITC-labeled antiserum against rabbit Ig (Behring), diluted 1:10. The washed smears were then mounted in

247

glycerine-PBS (9:1). 150 consecutive spermatozoa were observed for each sample on a Zeiss photomicroscope equipped with fluorescence and phase contrast at x480 magnification. The percentage of sperm with the occurrence of fluorescence over the anterior region of the head was evaluated in each sample.

Sopravital Staining. The percentage of alive sperm using eosin Y 0.1% was evaluated on samples of sperm suspensions before and after capacitation.

Transmission Electron Microscopy (TEM). Electron microscopical assessment of the acrosomal status was carried out on capacitated sperm samples from 2 fertile controls and 3 patients with negative SPA. Sperm samples (almost 3×10^6 spz) were fixed in PAF solution for 2-4 hours at 4°C and post-fixed in 1% OsO_4 solution in bidistilled water at 4°C for 1 h. The pellets of spermatozoa recovered upon centrifugation (1500 xg x 15 min.) were deh drated in graded alcohol and embedded in Epon 812. Thin sections were contrasted in uranyl acetate and lead hydroxide and examined on Zeiss 10CR electron microscope. In each sample almost 100 sperm head in longitudinal section were examined, and the proportion with intact acrosome or lacking an acrosome, with or without vesicular remnants, was determined.

RESULTS

SPA. SPA showed positive results (penetration rate $>$ 10%) in 41.6% of the patients. No relation was found between type of infertility and results of the test (TAB.I). All fertile patients showed positive results.

TAB.I. RESULTS OF SPA IN RELATION TO THE TYPE OF INFERTILITY

type of infertility	SPA + (n°cases)	SPA - (n° cases)
unexplained infertility	3	4
retrograde ejaculation	1	2
immunological infertility	0	1
asthenozoospermia	1	0

IIF. Table II shows the mean percentage of acrosin-positive spermatozoa before and after capacitation in three groups of subjects (patients with negative SPA, patients with positive SPA and fertile controls with positive SPA). The mean percentage of alive spermatozoa in the same groups before and after capacitation is also reported. The percentage of acrosin-positive sperm resulted neither significantly

248

different before and after capacitation in each group, nor significantly different from the percentage of alive sperm.

TAB.II. MEAN PERCENTAGE OF ALIVE AND ACROSIN-POSITIVE SPERM BEFORE AND AFTER CAPACITATION

PATIENTS	UNCAPACITATED SPERM		CAPACITATED SPERM	
	alive	acrosin+	alive	acrosin+
	means (%)+SD		means (%)+SD	
with negative SPA (n.7)	89+9.77	88.6+9.4	84.8+10.88	87.4+15
with positive SPA (n.5)	88+2.12	89.4+5.3	83.4+4.09	84.7+8.39
fertile controls	88.6+6.5	90.5+9.85	86 +8.09	85.6+9.54

TEM. The ultrastructural study of the 5 specimens of capacitated sperm confirmed that only a very low percentage of sperm heads did not show an intact acrosome. The percentage of intact acrosomes did not significantly differ from that of acrosin-positive sperm evaluated by IIF (Tab. III).

TAB.III. COMPARISON BETWEEN RESULTS OF TEM AND IIF TEST IN SPECIMENS OF CAPACITATED SPERM FROM 5 SUBJECTS

IIF TEST acrosin-positive mean (%)+ SD	TEAM intact acrosome mean (%)+ SD
78.4+7.95	80.8+3.9

Moreover, the sperm head without an intact acrosome usually showed morphological features of dead cells and only sporadically morphological features suggestive of "true" acrosomal reaction were observed.

DISCUSSION AND CONCLUSIONS

Our results suggest that "true" acrosome reaction does not represent a massive event during "in vitro" capacitation of human spermatozoa. This is in agreement with previous data obtained with IIF using an antiserum against acrosomal surface antigens (3) or TEM (4). Moreover, our data indicate that the results òf SPA are not in relation with the rate of acrosome reaction, which seems to represent a sporadic occurrence during "in vitro" capacitation, whatever SPA results may be.

REFERENCES

1) Rogers B.J. The sperm penetration assay: its usefulness reevaluated. Fertil. Steril. 1985: 43; 821-36
2) Yanagimachi R., Yanagimachi H., Roger B.J. The use of zona-free animal ova as a test system for the assessment of the fertilizing capacity of human spermatozoa. Biol. Reprod. 1976: 15; 471
3) Byrd W. and Wolf D.P. Acrosomal status in fresh and capacitated human ejaculated sperm. Biol. Reprod. 1986: 34; 859-869.
4) Nagae T., Yanagimachi R., Srivastava P.N. and Yanagimachi H. Acrosome reaction in human spermatozoa. Fertil. Steril. 1986: 45; 701-707
5) Flörke-Gerloff S., Töpfer-Petersen E., Miller-Esterl W., Mansouri A., Schatz R., Schirren C., Schill W. and Engel W. Biochemical and genetic investigations of round-headed spermatozoa in infertile men including two brothers and their father. Andrologia 1984: 16; 187-202

Use of acridine orange for morphological study on male infertility

G. GIORDANO-LANZA, S. MONTAGNANI, C. PAL-MA, M. DE ROSA[1], A. VITA[1], G. LOMBARDI[1]

Institute of Human Anatomy; [1]Institute of Endocrinological Science Chair of Endocrinology, 2nd School of Medicine, University of Naples, Naples, Italy

INTRODUCTION

The metachromatic fluorochrome acridine orange (AO) is widely used as a probe to investigate conformation of nucleic acid in situ: the dye fluoresces green when it interacts with double-stranded (ds) nucleic acids and exhibits red luminescence when it binds to single stranded (ss) polymers. Interaction of AO with ss nucleic acid (RNA and denatured DNA) forms an electrostatic binding with the phosphates of nucleic acid backbone which is responsible for the long wavelength red luminescence of the AO-ss-nucleic acid complexes (1, 2).
These complexes are precipitates insoluble at a wide range of ionic strength and it is possible to visualize them with electron microscopy (3). In addition, recent studies (4) suggested that AO may per se exert a denaturing effect on nucleic acids, because the dye binding to ss sections of DNA is strongly cooperative and a progressive denaturation of the adjacent ds sections could occur as a result of such binding.
We know that resistance of the sperm to heat denaturation of its DNA is an important parameter in fertility. According to Tejada (5,6) we use AO both as an active denaturing agent and as a fluorescent probe of DNA condition in four groups of infertile patients.

Materials and methods

We have studied 109 patients divided into four groups. The first group was of 29 patients affected with idiopathic hypogona-dotrophic hypogonadism (IHH) before and during treatment with gonado-tropins. The treatment had achieved complete spermatogenesis and normalization of testosterone plasma levels (7).

251

The 2nd group comprised 21 patients affected with varicocele (V) grade I and 2 according to Steeno (8) diagnosed by manual palpation, scrotal termography and eco-doppler fluxymetry.

The 3rd group comprised 35 subjects affected by idiopathic oligoasthenospermia (IOA), i.e. without any hormonal metabolic or systemic disease that could be imputable cause of it.

The 4th group comprised 34 patients affected by asymptomatic bacteriospermia (AB), confirmed by semen cultures and without any inflammation of genital tract. The fluorescent pattern of the patients was compared with that of 10 subjects of proven fertility. Hormonal and seminal patterns of four groups of patients are reported in Tab. I. Semen samples were obtained by masturbation and were studied only in the fresh state: they were washed within 30' of collections twice by centrifugation in sterile Tyrode's solution at 1300Xg for 5'. Smears were made on slides and allowed to air-dry for 30'; slides were then fixed in Carnoy, and re-dried at room temperature. Staining solution was daily prepared as follows:

10 ml of stock solution (1 g of AO in 1000 ml of distilled water)
40 ml of 0,1 citric acid
2,5 ml of 0,3M Na_2HPO_4. 7 H_2O.

Each slide was covered with staining solution for 5', then washed in several changes of distilled water and observed the same day on a fluorescence microscope.

A count of five fields in every slide with almost 300 cells gave our percentage of green cells. There was a good agreement among independent observers, with a standard deviation under 3%.

The percent of green cells has been multiplied by sperm count to obtain the "Effective Sperm Count" according to Tejada(5).

TAB.I:Hormonal and seminal parameters in 109 infertile patients

	N	Hormonal pattern			Seminal parameters		
		FSH	LH	T	N/mm^3	F.P.(%)	mot.1a h
IHH	29	5,81 ± 1,15	47,34 ± 12,88	678 ± 137,5	11.000 ± 2.100	30% ± 2,1	40% ± 5,5
V	21	8,5 ± 1,2	7,5 ± 1,1	515,1 ± 50,1	12.544 ± 4;202	35% ± 15	40% ± 25
IOA	35	8,9 ± 0,4	10,1 ± 0,8	491,5 ± 45,6	12.661 ± 2.071	30% ± 10	43% ± 20
AB	34	6,4 ± 0,3	9,2 ± 1	517,9 ± 36,1	15.835 ± 3.100	25% ± 10	35% ± 10

RESULTS

In every case,the control subjects exhibited a green sperm count of almost 50% ,while the patient's green values covered a much wider range.In fact,the patients of the 1st group exhibited 30-40% red coloured sperm heads.The patients affected with varicocele(2nd group) had 20-100% red cells.The 3rd group showed 40-90% red cells,while for the 4th the percentage was 30-90%.
The "Effective Sperm Count" confirmed defective fertility pattern for all the patients,while the control group had ESC values of 40millions or greater.As the ESC is the product of actual sperm count times the percentage of green cells,these two different values may influence the final result.So,we observed that particularly in the IHH patients,the low ESC was mostly due to low sperm number.In patients of 2nd and 4th group(i.e.V and AB)the % of green normal cells was very low.
In the subjects of the 3rd group(i.e.IOA),the effective sperm count was low for both the considered parameters.

DISCUSSION AND CONCLUSION

Our observations confirm the validity of AO-test according to Tejada to differentiate among normal and altered DNA content.Our results seem to display that fertility deficit,in pathologies not from uro - genital tract(i.e.IHH)is fundamentally due to sperm output.
On the contrary,in the groups with uro-genital noxae(i.e. V and AB), the percentage of red cells grows, most of all,for the direct action of bacteria or circulatory alterations on sperm DNA in the testis.
In OAI patients the effective green sperm cell count is low either for the decreased sperm production either for the low percentage of normal cells.In a recent study,Mitchell provided evidence that the red/green ratio is determined early in spermatogenesis and suggested that red sperm have poorer survival characteristic than the green.
Our patients generally had sperm counts lower than those of Tejada,so ESC put a lot of them in the subfertile/infertile region.These results are accorded with the infertility of almost all these patients,and provide additional information on their sperm DNA.
In conclusion,the use of AO in infertile patients with normal or abnormal ejaculates could be exploited to correlate sperm pathology with its morphology.AO staining is useful to demonstrate also bacteria or esfoliated cells,often present as causes or effects of the alterations of these ejaculates.

REFERENCES

1-Darzynkiewicz Z.,Evenson D.,Kapuscinski J.,Melamed M.R. Denaturation
of RNA and DNA in situ induced by Acridine Orange.Exp.Cell Res.,
148,31-46,1983.
2-Ichimura S. Differences in the red fluorescence of Acridine Orange
bound to single-stranded RNA and DNA.Biopolymers 14,1033-1047,1975.
3-Kapuscinski J.,Darzynkiewicz Z. Increased accessibility of bases in
DNA upon binding of Acridine Orange.Nucleic Acid Research 11,21,1983.
4-Hurst R.E.,Roy J.B. Acridine Orange male fertility test.Letter to
the Editor.Fertility and Sterility,43,1,1985.
5-Tejada R.I.,Cameron Mitchell J.,Norman A.,Marik J.J.,Friedman S.
A test for the pratical evaluation of male fertility by Acridine
Orange(AO)fluorescence.Fertility and Sterility,42,1,1984.
7-Steeno O.,Knops J.,Declerck L.,Adimoelija A.,Van de Voorde H.
Prevention of fertility disorders by detection and treatment of
Varicocele at school and college age.Andrologia,8,47-53,1976.
8-De Rosa M.,Lombardi G.,Torino G.,Quagliozzi L.,Aurigemma A.C.,Panza
N. Daily and alternated day treatment with an LH-RH analog in the
therapy of idiopathic hypogonadotropic hypogonadism.
J.Androl.,5,III,1984.

SUBJECT INDEX

255